Environmental Security in Transnational Contexts

W0113109

Much of the discussion surrounding the definition of the Sustainable Development Goals (SDGs) and the post-2015 global development agenda has contextualized sustainable development within the framework of 'transformation', specifically prioritizing concepts such as equity, security, justice, and rights. While these debates correctly discussed power imbalances and relational obstacles to human development they have remained abstract because they focused only on the international level. In this regard, discussions have not adequately examined mechanisms that facilitate or block the emergence of sustainable development as a political priority, nor do they address specific policy proposals to link environmental justice to human development strategies. This book contends that human and environmental security should be framed in terms of transnational discussions rather than being limited to general international debates in order to examine both governance challenges and potential policy mechanisms that can effectively address environmental security issues that cross national boundaries. The chapters in this volume undertake an empirical examination of the relationships between human and environmental security, cross-border exchanges, and regional integration. They address the relationships between international norms, transnational human and environmental security issues, and the regionalization of governance in different parts of the world as the book includes comparative analyses as well as case studies from Europe, Asia, and the Americas.

The book was originally published as a special issue of *Globalizations*.

Harlan Koff is Professor of Social Sciences at the University of Luxembourg. He is President of the Consortium for Comparative Research on Regional Integration and Social Cohesion (RISC) (www.risc.lu) and co-editor of the journal *Regions & Cohesion* (Berghahn Journals). He conducts research on migration, international development, regional integration, and borderlands. He is Docent in Development Studies at the University of Helsinki, Finland.

Carmen Maganda (PhD) is Research Professor of Environment and Sustainability at the Instituto de Ecología (INECOL), Mexico. She is co-editor of the journal *Regions & Cohesion* (Berghahn Journals). Her research focuses on social participation in environmental governance, water security/justice, and regional integration.

Rethinking Globalizations

Edited by Barry K. Gills, *University of Helsinki, Finland* and
Kevin Gray, *University of Sussex, UK.*

This series is designed to break new ground in the literature on globalization and its academic and popular understanding. Rather than perpetuating or simply reacting to the economic understanding of globalization, this series seeks to capture the term and broaden its meaning to encompass a wide range of issues and disciplines and convey a sense of alternative possibilities for the future.

For more information, please visit:
https://www.routledge.com/Rethinking-Globalizations/book-series/RG

Environmental Security in Transnational Contexts

What Relevance for Regional Human Security Regimes?

Edited by
Harlan Koff and Carmen Maganda

LONDON AND NEW YORK

First published 2018
by Routledge

2 Park Square, Milton Park, Abingdon, Oxfordshire OX14 4RN
52 Vanderbilt Avenue, New York, NY 10017

Routledge is an imprint of the Taylor & Francis Group, an informa business

First issued in paperback 2020

British Library Cataloguing in Publication Data
A catalogue record for this book is available from the British Library

ISBN 13: 978-0-8153-8598-1 (hbk)
ISBN 13: 978-0-367-51816-5 (pbk)

Typeset in Times New Roman
by RefineCatch Limited, Bungay, Suffolk

Publisher's Note
The publisher accepts responsibility for any inconsistencies that may have
arisen during the conversion of this book from journal articles to book chapters,
namely the possible inclusion of journal terminology.

Disclaimer
Every effort has been made to contact copyright holders for their permission to
reprint material in this book. The publishers would be grateful to hear from any
copyright holder who is not here acknowledged and will undertake to rectify
any errors or omissions in future editions of this book.

Contents

Citation Information

The chapters in this book were originally published in *Globalizations*, volume 13, issue 6 (December 2016). When citing this material, please use the original page numbering for each article, as follows:

Introduction
Environmental Security in Transnational Contexts: What Relevance for Regional Human Security Regimes?
Harlan Koff & Carmen Maganda
Globalizations, volume 13, issue 6 (December 2016), pp. 653–663

Chapter 1
Reconciling Competing Globalizations through Regionalisms? Environmental Security in the Framework of Expanding Security Norms and Narrowing Security Policies
Harlan Koff
Globalizations, volume 13, issue 6 (December 2016), pp. 664–682

Chapter 2
Water Security Debates in 'Safe' Water Security Frameworks: Moving Beyond the Limits of Scarcity
Carmen Maganda
Globalizations, volume 13, issue 6 (December 2016), pp. 683–701

Chapter 3
Scarcity and Power in US–Mexico Transboundary Water Governance: Has the Architecture Changed since NAFTA?
Stephen P. Mumme
Globalizations, volume 13, issue 6 (December 2016), pp. 702–718

Chapter 4
Many Faces of Security: Discursive Framing in Cross-border Natural Resource Governance in the Mekong River Commission
Andrea K. Gerlak & Farhad Mukhtarov
Globalizations, volume 13, issue 6 (December 2016), pp. 719–740

For any permission-related enquiries please visit:
http://www.tandfonline.com/page/help/permissions

Notes on Contributors

Andrea K. Gerlak holds a joint position at the University of Arizona, USA, as Associate Professor in the School of Geography and Development and as Associate Research Professor at the Udall Center for Studies in Public Policy. Her work focuses on how we can better design institutions to promote adaptive, flexible policies to improve human and ecosystem well-being, and produce fair and equitable decisions.

Sylvain Guyot is a Geographer and Full Professor at the University of Bordeaux Montaigne, UMR 5319 CNRS, junior member of IUF (Institut Universitaire de France), and Director of the Territorial Innovation Master's program. His research focus is on art and nature in world-wide contexts.

Harlan Koff is Professor of Social Sciences at the University of Luxembourg. He is President of the Consortium for Comparative Research on Regional Integration and Social Cohesion (RISC) (www.risc.lu) and co-editor of the journal *Regions & Cohesion* (Berghahn Journals). He conducts research on migration, international development, regional integration, and borderlands. He is Docent in Development Studies at the University of Helsinki, Finland.

Carmen Maganda (PhD) is Research Professor of Environment and Sustainability at the Instituto de Ecología (INECOL), Mexico. She is co-editor of the journal *Regions & Cohesion* (Berghahn Journals). Her research focuses on social participation in environmental governance, water security/justice, and regional integration.

Farhad Mukhtarov is a Research Fellow at Copernicus Institute of Sustainable Development, Utrecht University, The Netherlands and Adjunct Senior Research Fellow at the Institute of Water Policy, National University of Singapore. Farhad's research focuses on how and to what effect policy models travel across various boundaries, and how global narratives translate into local practices. He is fascinated by the challenges of institutional design for sustainability under conditions of contingency and contextuality.

Stephen P. Mumme is Professor of Political Science at Colorado State University, USA, where he specializes in comparative environmental policy with an emphasis on water and environmental management along the US border with Mexico.

Bastien Sepúlveda received his PhD in Geography from the Université de Rouen in 2011. He is currently a Research Fellow at the Université de Lille, where he leads the ACHN project "Indigenous geographies: roots, developments and perspectives in the French-speaking universe", funded by the French Agence Nationale de la Recherche.

NOTES ON CONTRIBUTORS

Jeroen Frank Warner is Associate Professor of Disaster Studies at Wageningen University, The Netherlands, where he also earned his PhD degree. His main research interests in the disaster studies domain are social resilience, the politics of (flood) disaster risk reduction, and the role of disaster in International Relations. He was Visiting Professor at the University of Sao Paulo in Autumn 2013 and is currently Special Visiting Professor at this institution.

INTRODUCTION

Environmental Security in Transnational Contexts: What Relevance for Regional Human Security Regimes?

HARLAN KOFF & CARMEN MAGANDA

ABSTRACT   *The year 2015 was meant to be a seminal year in global geopolitics due to the transition from the Millennium Development Goals (MDGs) to the Sustainable Development Goals (SDGs). This transition was significant because the MDGs, even though they raised global consciousness around the need to combat poverty, remained indicator-based, and thus, they did not adequately address socio-economic inequalities and power imbalances in global affairs. For this reason, much of the discussion surrounding the definition of the SDGs and the post-2015 global development agenda contextualized sustainable development within the framework of 'transformation', specifically prioritizing concepts such as equity, security, justice, and rights. While these debates correctly discussed power imbalances and relational obstacles to human development they remained abstract because they focused only on the international level. In this regard, discussions did not adequately examine mechanisms that facilitate or block the emergence of sustainable development as a political priority, nor did they address specific policy proposals to link environmental justice to human development strategies. Thus, this special issue introduction argues that human and environmental security should be framed in terms of transnational discussions rather than being limited to international debates. The special issue undertakes an examination of the interactions between human and environmental security, border studies, and comparative regional integration; and interactions between competing globalizations. The articles in the special issue address the relationships between international norms, transnational human and environmental security issues, and the regionalization of governance.*

Introduction

The year 2015 was meant to be a seminal year in global geopolitics due to the transition from the Millennium Development Goals (MDGs) to the Sustainable Development Goals (SDGs). In short, this transition is significant because the MDGs, even though they raised global conscious-ness around the need to combat poverty, remained indicator-based, and thus, they did not ade-quately address socio-economic inequalities and power imbalances in global affairs. Conversely, the proposed SDGs are expanding the proposed number of objectives to be addressed in the post-2015 global development agenda from 8 to 17 and they include both sector-specific objectives such as SDG 6 ('Ensure availability and sustainable management of water and sanitation for all') to be addressed within nation-states and norm-based goals such as SDG 10 ('Reduce inequality within and among countries') and SDG 16 ('Promote peaceful and inclusive societies for sus-tainable development, provide access to justice for all and build effective, accountable, and inclusive institutions at all levels') that also address relations between nation-states.

Much discussion has already taken place in international affairs in relation to the definition of the SDGs and the post-2015 global development agenda. Building on criticisms of the MDGs' focus on indicators, many observers have contextualized sustainable development within the fra-mework of 'transformation', specifically prioritizing concepts such as equity, security, justice, and rights (see Institute of Development Studies, 2015). For example, Jeffrey Sachs, Special UN Advisor on the MDGs has identified the need to move beyond poverty-reduction to a 'triple bottom line' approach that includes economic development, environmental sustainability, and social inclusion (Sachs, 2012). This approach has been echoed by other observers, such as Battacharya, Khan and Salma (2014) who included in their comment on the report of the UN Open Working Group on SDGs their opinion that the document has not significantly addressed 'a consensus on transformative development' that identifies which aspects of the SDGs are 'uni-versal' and which ones are 'national'. Martens (2015) has also indicated that the SDGs must embrace a 'universal' approach to development in order to promote real change in development cooperation relationships that address power imbalances in international economic and financial systems.

While these debates correctly identify power imbalances and obstacles to human develop-ment, defined as development aimed at maximizing human capabilities (see United Nations Human Development Report, 1994), they remain abstract because they focus only on the inter-national level. In this regard, discussions do not adequately examine mechanisms that facilitate or block the emergence of sustainable development as a political priority, nor do they address specific policy proposals to link environmental justice to human development strategies.

This special issue, which results from a writers' workshop sponsored by the Consortium for Comparative Research on Regional Integration and Social Cohesion (RISC) and the University of Luxembourg funded HUMENITY (Human and Environmental Security in Border Regions: Cross-regional Perspectives) research project, contends that international approaches to sustain-able development strategies, linking environmental, and human security discussions need to be grounded in transnational contexts defined as local frameworks that cross nation-state borders. It makes these arguments for two reasons: (1) threats to environmental security often cross borders and for this reason, domestic policies are no longer sufficient to address these threats. Bilateral and multilateral cooperation is a necessity today in transnational environmental governance aimed at resolving specific environmental problems. Moreover, inter-state competition can con-tribute to the exacerbation of threats to environmental security; (2) throughout the world, the proliferation of regional integration has added an important level of governance in the field of human security which is often ignored. Even though degrees of region-building vary significantly, every world region has witnessed in one way or another, the cession of sovereignty by nation-states to regional bodies in determined policy arenas that affect human security.

For these reasons, this special issue argues that human and environmental security should be framed in terms of *transnational* discussions rather than being limited to *international* debates in response to the research question 'How does cross-border environmental security relate to international discussions of human security, such as those that characterize the 2015 SDG debates?' The special issue focuses on the emergence of the human and environmental security paradigms internationally and above all, their application (or lack thereof) at the local, cross-border level. Specifically, the special issue identifies regional human security regimes as important vehicles for the transmission of international norms to transnational environmental security strategies. In doing so, its originality is the comparative examination of the interactions between human and environmental security, border studies, and comparative regional integration, three fields of research which have rarely been linked despite their relevance to each other. These connections are significant because human and environmental security are not only objective measures of risks/threats to human well-being, they are subjective policy concepts that are socially/politically constructed through competing globalizations (e.g. international organizations, transnational social movements, cross-border institutions, etc.). The interconnections between these globalizations are the focus of this special issue.

The Emergence of Human and Environmental Security

Security politics have changed radically since the end of the Cold War. Previously, this concept had been defined in state-terms as the national security paradigm placed countries squarely at the center of analyses and threats were viewed militarily. Since the end of the Cold War, however, our frameworks of analysis have changed significantly as two new security paradigms emerged which focused on individuals rather than states: Human Security and Environmental Security. These concepts, which derive from the United Nations Development Programme's (UNDP) milestone 1994 Human Development Report and the 1987 Brundtland Report, famously discuss 'freedom from want' and 'freedom from fear' as benchmarks for human security and 'development that meets the needs of the present without compromising the ability of future generations to meet their own needs' as the basis for environmental security (as well as the definition of sustainable development). While the declaration of these concepts indicated an important paradigm shift, critics rightfully argue that the vagueness of these ideas has hindered their effectiveness. Moreover, important debates have arisen focusing on whether/how environmental security should be embedded in human security approaches. These questions are central to the post-2015 SDG Agenda where leaders aim to strike a delicate balance between security, economic development, environmental conservation, social justice and governance.

Security from What? Security for Whom? . . . It is all in the Definitions

Politics, of course is about agenda-setting and conceptual definitions significantly affect policy strategies. In this regard, the emergence of human security and environmental security represent important developments in global affairs. First and foremost, these concepts broke state monopolies on security issues. Threats were no longer defined in state-terms and the protection of citizens was no longer assigned only to governments.

Classical theories of state-formation (see Tilly, 1993) have contended that citizens expect two things from their states: (1) protection and (2) the just provision of common goods and services. It is also accurate to note that in international relations, we have traditionally defined nation-states institutionally (see Grieco, 1988) as governments have been charged with providing the aforementioned protection and services when discussing global security. Instead, the human security and environmental security paradigms return security debates to classical political science definitions of the 'State' which is viewed as an organized political community that lives under a system of government (including identifiable rights and responsibilities) within an identifiable territory (see Almond, 1988). Under this definition, the focus of security shifts away from governments and reintroduces citizens as both the agents and subjects of security politics (see Gomez, 2015).

Of course, 'bringing the people back in' to security debates has democratized discussions and led to more representative processes. This explains the differences between the government negotiations that led to the more restrictive MDGs and the participative consultation process known as 'The World We Want' which has preceded the proposed SDGs and indicated that they will be more transformative in nature. However, this conceptual strength of human security and environmental security paradigms has also been its weakness: international policy-making in these fields has rarely surpassed the stage of summitry.

'Sustainable Development', 'Human Security', and 'Environmental Security' are all concepts that gained prominence in the 1980s through international summits. During this period, international organizations such as the United Nations (UN) and the European Union (EU) began breaking with traditional notions of development that focused on financial transfers between rich and poor countries. Many development treaties, such as the European Union's Lomé Conventions directly resulted from colonial traditions (the Lomé Conventions were championed in the European agenda by the French). In a break from this approach, The International Union for the Conservation of Nature and Natural Resources (IUCN) presented sustainable development as the objective of its World Conservation Strategy. The position that was forwarded shifted the focus of development from economics to ecology and 'Environmental Security' emerged.

In 1987, the aforementioned Brundtland Report was published under the title *Our Common Future* which oriented security debates toward the future rather than the (colonial) past. This was followed in 1992 by the UN Conference on Environment and Development's Earth Charter which highlighted the need to build a just, sustainable, and peaceful society. This document, and the ensuing *Agenda 21* action plan, infused the notion of sustainable development with important socio-political themes. The concept of 'sustainability' was broadened and linked to the notions of human rights, human security, environmental conservation, the protection of indigenous peoples, etc. It has become virtually impossible to politically oppose sustainable development because of its multifaceted, complex nature which includes attractive themes for a diverse group of proponents.

This process culminated in the establishment of the MDGs in 2000. In many ways, the Millennium Development Summit, held at the United Nations in New York, represents the apex of global affairs. Following the establishment of eight MDGs, all 189 United Nations member states at the time committed to help achieve these goals. This historical moment represents the first declared global commitment to the eradication of poverty and it signified the inclusion of environmental security and human security into the universal global political consciousness on development.

Since 2000, however, sustainable development discussions have been faced with considerable difficulties. The summits that were championed as political endorsements of sustainability and progressive steps forward in the 1990s, often ended in frustration in the 2000s. For example (and most visibly), the 2009 United Nations Climate Change Conference (the Copenhagen Conference) was characterized by disarray, in-fighting, and political disagreement between the United States and its allies and the BRICS (Brazil, Russia, South Africa, India, and China). Similarly, the 2012 Rio + 20 Summit illustrated the limits of consensus in global sustainable development debates. The summit achieved renewed support from world leaders for sustainable development as a political objective. However, many participants were frustrated by the lack of operationalized commitments toward sustainability. While 'sustainable development' is not synonymous with 'human security' and 'environmental security', it does encompass these concepts in a vision of development. For this reason, one could say that the SDGs are premised by human and environmental security. While the declaration of these goals does represent a step forward in defining human and environmental security, it is yet to be seen whether it will actually lead to operationalized policy strategies. Given this inherent weakness in international discussions of human and environmental security, this special issue contends that transnational approaches should be examined as a means to implement these norms.

Human and Environmental Security in Transnational Contexts

Threats to security have become one of the most prominent themes in border politics throughout the world. Issues such as unauthorized migration, drug trafficking, human trafficking, arms trafficking, etc. have all become associated with contemporary border debates. In fact, scholars such as Andreas (2000), Brunet-Jailly (2007), and Sabet (2013) have noted that borders now occupy center stage in most national security discussions. This is also evident in the important increases in the budgets of border control agencies such as FRONTEX (European Union) and Homeland Security (United States). This scholarship also notes that globalizations have re-defined security threats from issues concerning the protection of the integrity of territories to questions related to the management of globalized flows.

These flows, in fact, are the forces that led to the emergence of human and environmental security in localized transnational security discussions. Whereas previous security paradigms aimed at 'securing a country's borders', the presence of flows have forced local authorities to collaborate with their counterparts on the other side of national divides in order to combat common security threats. Cornelius (2004), Koff (2008), and Ceballos (2011) have empirically demonstrated that the incentive structures of local authorities in regard to security issues often contradict those of national officials. Stated more directly, local border communities are often the victims of collateral damage of centralized decision-making in the field of security (see Cortez-Lara, 2012; Maganda, 2005). When central governments close and or militarize borders, the socio-economic and environmental costs of these decisions are often quite high for border populations.

Consequently, whereas international security debates often remain conceptual and normative in nature, transnational security discussions are generally issue-based. The rich literature on borderlands studies has noted that threats related to human and environmental security have emerged as political priorities for cross-border communities where the impacts are felt on a daily basis. These include issues such as: water security, environmental risk and disasters, land and air pollution, organized crime, unauthorized migration, contraband and trafficking of illegal goods, etc. Moreover, much scholarship in the fields of borderlands studies (see Bruns et al., 2013; Kopinak, 2004; Peberdy, 2000; Staudt, 1998) has posited that the liberalization of regional markets through economic globalization has contributed to social marginalization and the rise of informal economies as threats to human and environmental security in many cross-border areas (see Koff & Maganda, 2014). Moreover, these socio-economic forces have contributed to social unrest and citizen dissatisfaction in many border regions. For this reason, scholars such as Staudt (2008), Simpson (2013), etc. have documented increased social mobilization amongst citizens of border communities against national authorities on human and environmental security issues. In parts of the world where government representatives and state structures are weak or illegitimate, these forms of social mobilization have been converted into rebellion which has further exacerbated transnational security problems in those areas (see Syria/Iraq; Mali; Pakistan/Afghanistan, etc.). These border regions are in fact, often characterized by informal structures of authority that are more powerful and sometimes more legitimate locally than formal state structures.

This specific trait of transnational human and environmental security discussions is also a defining characteristic: because representative cross-border security structures are rare, most discussions in this field are informal in nature. While decision-making structures do exist in some policy arenas, such as cross-border water councils or economic development and planning bodies, etc. security is only one policy issue discussed within broader governance systems and as such, it is not prioritized. For this reason, observers such as Payan (2006) have noted that cross-border security discussions are defined by an inherent democratic deficit. Paradoxically, whereas international security debates are becoming more consultative, localized cross-border security discussions are evolving in the opposite direction as decisions are being made behind closed doors. This is leading to political tensions within border communities where human and environmental security are serious political issues that affect the quality of life of local inhabitants. What could be an answer to this paradox? Some observers (see Kirchner & Dominguez, 2011; van Langenhove et al., 2009) have proposed regional responses to human and environmental security problems. This is the subject of the following section.

Regional Integration and Human and Environmental Security: Caught Between a Rock and a Hard Place?

Harlan Koff's contribution to this special issue has cited the emergence of the literature on regional security regimes and their relationship to both human and environmental security issues. Fawcett (2013) has noted that 'security regionalisms' have become prominent in specific security fields (e.g. border controls, migration policies, anti-terrorism, etc.) but that, in general, these organizations have not developed strong regional security institutions. Similarly, van Langenhove et al. (2009) have recognized the significant contributions that numerous regional organizations have made to the protection of human security in different world regions. They too, however, have contended that the successful implementation of human security policies depends on multilevel governance and coordinated actions at the supranational, national, and sub-national levels.

These studies, amongst others, provide clues as to why environmental security has not figured prominently in regional security regimes. They have indicated that unlike other policy arenas where regions have taken leading roles in policy implementation, security has remained an issue dominated by member states which are hesitant to cede authority to regional organizations over hard security questions. This is evident in the European Union where Common Foreign and Security Policy (CFSP) has been one of the policy arenas in which supranationalism has been slowest to develop (see Hix & Hoyland, 2011).

The most complete comparative study in this field has been conducted by Kirchner and Dominguez (2013). These authors examined the relationship between regional security performance and the domestic economic and political conditions of 14 regional organizations' member states. In general, the authors of this study noted that member states have readily supported regional organizations in the definition of regional security policies. However, their support has waned through the policy coordination and implementation stages. Consequently, member states supported regional security discourse without necessarily supporting operationalized regional security policies.

These studies, amongst other, have documented institutionalized difficulties for the implementation of human and environmental security at the regional level. In fact, despite the fact that regional organizations seem to be well placed in the international system of governance to address both international and transnational human and environmental security questions, these bodies suffer from important strategic weaknesses that limit their actions in these fields. First, it must be noted that most regional organizations do not participate in the UN system where human and environmental security norms are defined. The EU was invited to participate in the 2015 SDG Summit as an observer but non-UN regional bodies generally cannot actively participate in the discussions that define international human and environmental security norms. Borrowing from the literature on policy coherence for development, defined as policy measures that aim to assure that development policies are not undermined by actions in non-development policy spheres, this situation can be considered an important inter-institutional incoherence because, by excluding regional organizations from the definitions of norms, the international system is blocking an important policy exchange mechanism. Regional organizations have less incentive to implement human and environmental security policies because they do not appropriate them as their own from the beginning.

Of course, the member states of these regional bodies are present at international summits where human and environmental security norms are defined. In theory, they would upload these norms to the regional level. However, this process is too closely related to power-sharing issues in regionalism. Scholars such as Kuhnhardt (2010) have noted that states are reluctant to cede sovereignty to regional bodies. Even though regions are well placed to link international human and environmental security norms to concrete transnational security strategies, the cession of sovereignty in 'soft' security domains could be viewed as a threat to sovereignty on 'hard' security issues (such as national defense) through policy spillover. For this reason, member states are reluctant to reinforce regional architectures in this policy arena and commitments to human and environmental security remain discursive in nature.

Finally, it must be noted that human and environmental security strategies are not necessarily implemented by regional organizations because of the important democratic deficits that characterize many of these bodies. This introduction has already indicated that human security and environmental security are policy objectives that have been championed by non-governmental organizations and they have emerged in a policy-making process in the field of international development that is increasingly becoming characterized by public consultation and civil society representation. Unfortunately, this process has not yet been replicated by many regional organizations. On one hand, many regional bodies are not inclusive as effective environmental governance structures are lacking and policy-making is characterized by a dearth of public information on the access to natural resources (see Maganda, 2008). However, it is also true that many member states do not foster public participation in regional policy-making because this indirectly weakens the position of nation-states within regions by directly legitimizing the authority of regional bodies in specific policy arenas, such as security. Consequently, even though regions are well placed structurally to reconcile the international and transnational components of human and environmental security, these deficiencies that presently characterize regional governance in many parts of the world must be addressed before regional organizations can realize their policy-making potential. The articles included in this special issue address these relationships between international norms, transnational issues, and regional governance.

Presentation of the Special Issue

As stated above, this special issue examines the complex relationships between international norms, transnational human and environmental security debates, and the regionalization of governance. The articles presented here aim to contribute to human and environmental security debates by examining these different elements of security politics as well as the mechanisms that link them. Another added value of the special issue is its focus on different world regions. The articles that follow include comparative regional analysis as well as case studies on human and environmental security issues in Europe, Asia, and the Americas. By varying the regional cases, the special issue accounts for how human, and environmental security norms are contextualized differently in distinct parts of the world within the framework of comparative analysis. This is one of the originalities of the research promoted by the RISC Consortium (www.risc.lu).

The first part of the special issue directly addresses international norms in the fields of human security and environmental security. The lead article, by Harlan Koff, comparatively examines environmental security regimes in 16 regional organizations and asks whether regions can effectively implement international environmental security norms. The article highlights the differences between international and transnational approaches to human and environmental security and it discusses how policy implementation amongst regional organizations reflects the character of their own regional security architectures more than normative commitments to environmental justice. The article contends that the emergence of regional environmental security regimes should be fostered by reinforcing regional security architectures through public participation mechanisms.

This is followed by a contribution from Carmen Maganda that questions the meaning of international norms in the field of environmental security. Maganda asks, 'Why is it relevant to discuss water security issues in cases where water availability and accessibility do not seem to be a problem?' In her response to this query, the author contends that international norms, in order to be universal, must be supported in states where they may not directly benefit from those norms. This human rights logic asks: What value do human rights have if they are not implemented universally? Empirically, Maganda operationalizes this conceptual approach through the study of the Human Right to Water and Sanitation in the European Union and particularly, in the Grand Duchy of Luxembourg, a small, wealthy state that is characterized by water surplus. The article questions whether environmental security has indeed emerged as an international norm given that resistance has characterized the responses of EU and Luxembourgish water officials to human rights approaches to water management.

The second grouping of articles included in this special issue focuses specifically on policy relationships between regional governance and environmental security. The contribution by Stephen Mumme examines the politics of water allocation on the United States-Mexico border since the implementation of the North American Free Trade Agreement (NAFTA) in 1994. The article contends that while NAFTA reforms have modestly changed the water allocation regime, they have not altered longstanding power relationships governing the allocation of water resources between the two countries. The case evidence suggests that theoretical constructs related to environmental security and regional governance, such as multilevel governance and collaborative watershed management, etc., are often overshadowed/undermined by more comprehensive power-sharing relationships within regional regimes.

The article by Andrea Gerlak and Farhad Mukhtarov also opens the black box of regional governance in the field of environmental security. This contribution examines how security is framed in the context of international river basin organizations (RBOs). RBOs are key regional organizations in transboundary water governance and operate in many international river basins around the world. As an example of cross-border governance, RBOs can promote joint cooperation and information sharing, and serve as a forum to bring together diverse stakeholders, thus reinforcing environmental security. The article examines how diverse actors frame security in the context of RBOs and at various scales and around certain management actions in a case study of the Mekong River Commission. Particular attention is paid to the links between water security, food security, and energy security in the broader water and development discourse, thus linking environmental security to human security issue arenas.

These transnational security issue arenas are the focus of the third grouping of articles included in this special issue. Jeroen Warner's contribution discusses whether linkage promotes transnational cooperation and political integration in cross-border water management. Linkage is conceptualized through a juxtaposition that is highlighted by the author through a typology that asks whether linkages between actors or issues are simply (but irrevocably) there, such as geographical proximity, or if they need to be deliberatively, artificially connected through an intervention to establish linkage that previously was not there. Warner applies this idea to the transboundary lower river Meuse and 'finds its history of integration to be a tortuous one' that is not linear and at various times has been characterized by cooperation and conflict which has affected both linkages and de-linkages. The author contends that transnational linkages in water security have not established broader regional integration in the field of environmental security through spillover.

The final article presented here, by Bastien Sepulveda and Sylvain Guyot, focuses on the management of protected areas in transboundary contexts and empirically examines the contemporary evolution of the border between Chile and Argentina in Northern Patagonia, which is a region that has witnessed the creation of numerous protected areas that are currently claimed by Mapuche organizations and communities as part of their customary territory. In response to these claims, both states have progressively integrated Mapuche communities into the management of protected areas through specific agreements. A new environmental governance model that includes the protection of indigenous peoples' rights is under construction not only along but also across the border between Chile and Argentina. Therefore, the authors contend that participatory management could be viewed as a tool for redefining borders through policy spillover by linking environmental security in protected areas to human security issues, such as the protection of ethnic groups' rights. Such cross-border participative approaches, of course, ground the international norms described above related to human security and environmental security in tangible transnational policy strategies and they also strengthen the institutional bases for regional governance, thus closing the analytical circle presented in this special issue and suggesting potential avenues for future progress.

Disclosure Statement

No potential conflict of interest was reported by the authors.

References

Almond, G. (1988). The return to the state. *The American Political Science Review*, 82(3), 853–874.

Andreas, P. (2000). *Border games: Policing the U.S. Mexico divide*. Ithaca, NY: Cornell University Press.

Battacharya, D., Khan, T. I., & Salma, U. (2014). A commentary on the final outcome document of the open working group on SDGs. *SAIS Review of International Affairs*, 34(2), 165–177.

Brunet-Jailly, E. (2007). *Borderlands*. Ottawa: University of Ottawa Press.

Bruns, B., Miggelbrink, J., Belina, B., Müller, K., Wust, A., & Zichner, H. (2013). Making a living on the edges of a security border: Everyday tactics and strategies at the eastern border of the European Union. In P. Gilles, H. Koff, C. Maganda, & C. Schulz (Eds.), *Theorizing borders through analyses of power relationships* (pp. 109–131). Brussels: P.I.E.-Peter Lang.

Ceballos Medina, M. (2011). La politica migratoria de Ecuador hacia Colombia: Entre la integración y la "contención". *Regions & Cohesion*, 1(2), 45–77.

Cornelius, W. (2004). Death at the border: Efficacy and unintended consequences of US Immigration Control Policy. *Population and Development Review*, 27(4), 661–685.

Cortez-Lara, A. (2012). No longer strong social cohesion: Lessons from two transboundary water conflicts in the Mexicali Valley, México. *Regions & Cohesion*, 2(2), 30–56.

Fawcett, L. (2013). *Security regionalisms: Lessons from around the world* (RSCAS Working Paper 2013/62). Fiesole: European University Institute.

Gomez, O. (2015). Visones alternativas sobre seguridad en América Latina: una contribución global en seguridad humana. *Regions & Cohesion*, 5(1), 26–53.

Grieco, J. (1988). Anarchy and the limits of cooperation: A realist critique of the newest liberal institutionalism. *International Organization*, 42, 485–508.

Hix, S., & Hoyland, B. (2011). *The political system of the European union* (3rd ed.). New York, NY: Palgrave Macmillan.

Institute of Development Studies. (2015). *Sustainable development goals must consider security, justice and equality to achieve social justice* (IDS Policy Briefing 88). Brighton: IDS.

Kirchner, E., & Dominguez, R. (2011). *The security governance of regional organizations*. London: Routledge.

Koff, H. (2008). El poder político y la política fronteriza en Europa: la utilidad de comparar las fronteras internas y externas de la UE. *Estudios Políticos*, 32, 195–226.

Koff, H., & Maganda, C. (2014). Water security in cross-border regions: What relevance for federal human security regimes? In D. Garrick, G. Anderson, D. Connell, & J. Pittock (Eds.), *Federal rivers: Managing water in multilayered political systems* (pp. 325–338). Cheltenham: Edward Elgar Publishing.

Kopinak, K. (2004). *The social costs of industrial growth in Northern Mexico*. La Jolla: Center for US-Mexico Studies.

Kuhnhardt, L. (2010). *Region-building: The global proliferation of regional integration* (Vol. I). New York, NY: Berghahn Books.

Maganda, C. (2005). Collateral damage: How the San Diego-imperial valley water agreement affects the Mexican side of the border. *The Journal of Environment & Development*, 14(4), 486–506.

Maganda, C. (2008). Agua dividida, agua compartida? Acuiferos transfronterizos en Sudamérica, una aproximación. *Estudios Políticos*, 32, 171–194.

Martens, J. (2015). Benchmarks for a truly universal post-2015 agenda for sustainable development. *Regions & Cohesion*, 5(1), 73–94.

Payan, T. (2006). *The three U.S.–Mexico border wars: Drugs, immigration, and homeland security*. Santa Barbara, CA: Praeger Security International.

Peberdy, S. (2000). Border crossings: Small entrepreneurs and cross-border trade between South Africa and Mozambique. *Tijdschrift voor Economische en Sociale Geografie*, 91(4), 361–378.

Sabet, D. (2013). Border burden: Public security in Mexican border communities and the challenge of polycentricity. In P. Gilles, H. Koff, C. Maganda & C. Schulz (Eds.), *Theorizing borders through analyses of power relationships* (pp. 79–103). Brussels: P.I.E.-Peter Lang.

Sachs, J. (2012). From millennium development goals to sustainable development goals. *The Lancet*, 379(9832), 2206–2211.

Simpson, A. (2013). Challenging hydropower development in Myanmar (Burma): Cross-border activism under a regime in transition. *The Pacific Review*, 26(2), 129–152.

Staudt, K. (1998). *Free trade? Informal economies at the US–Mexico border*. Philadelphia, PA: Temple University.

Staudt, K. (2008). *Violence and activism at the border*. Austin: University of Texas Press.

Tilly, C. (1993). *Coercion, capital and European states: AD 990–1992*. Hoboken, NJ: Wiley-Blackwell.

United Nations Human Development Report. (1994). *Redefining security*. New York: United Nations Secretariat.

van Langenhove, L., Vigilante, A., Fanta, E., Felício, T., Ferro, M., Scaramagli, T., & Tavares, R. (2009). *The regional dimension of human security: Lessons from the European Union and other regional organizations*. Garnet Policy Brief 9.

Reconciling Competing Globalizations through Regionalisms? Environmental Security in the Framework of Expanding Security Norms and Narrowing Security Policies

HARLAN KOFF

ABSTRACT *This article examines environmental security regimes in 16 regional organizations and asks whether regions can effectively implement environmental security norms. It first defines these norms and discusses their emergence at the international level. At the same time, through the literature review, the article posits that the globalization of security threats has simultaneously led to a retrenchment of coercive non-state security strategies. Consequently, the article contends that the globalization of security norms has made them ineffectual because they have not properly addressed tangible security threats. At the same time, nation-state-based hard power security measures (especially border controls) have not adequately addressed the underlying causes of transnational threats related to human and environmental security. For this reason, the article examines how well regional approaches to security contribute to both protection against imminent violence and the promotion of human and environmental security through medium-term development strategies. The article contends that the emergence of regional environmental security regimes should be fostered by reinforcing regional security architectures through public participation mechanisms.*

1. Introduction

Since the end of the Cold War, security debates have become quite nebulous. Internationally, 'new' global security norms such as human security, environmental security, water security, climate security, responsibility to protect (R2P) have emerged. Conversely, nation-state-based security policies have become narrow with specific concentrations on terrorism, organized crime, trafficking, migration, border control, etc. This article addresses this seeming disconnect between international discourse and domestic strategies through a comparative analysis of regional security regimes. The article addresses three specific research questions: (1) Can regional organizations reconcile the inconsistencies between international security norms and domestic policies with regard to environmental security? (2) Why is environmental security not prominent in regional security agendas?, and (3) What is the impact of inter-regionalism on environmental security policies? It contends that regions are better positioned than nation-states to address environmental security issues because they can coordinate/implement transnational security strategies. However, the potential for regions to do so has been muted due to weak security architectures.

Following this introduction, the article includes four sections and a conclusion. In Section 2, the article presents a literature review and posits that the globalization of security norms has made them ineffectual at the international level because they have not properly addressed tangible security threats through soft power mechanisms. At the same time, nation-state-based hard power security measures (especially border controls) have not adequately addressed the underlying causes of security threats related to poverty, global inequalities, socio-economic and environmental vulnerability, etc. For this reason, the article argues that regional approaches to security can combine, when member states are committed, hard power responses to immediate threats and soft power strategies to the underlying transnational causes of insecurity. Consequently, regions can contribute to protection against imminent violence and they can address human and environmental security issues through medium-term development strategies.

In Sections 3 and 4, the article responds to the first two research questions above by comparing the security agendas of 16 world regions and the places that environmental issues have in them. According to the literature on regional security, there has been an increase in the variance between world regions in how they define security and how they address environmental issues (see Fawcett, 2013). This section documents the differences between the security discourses of regional organizations and specifically, it examines how environmental security figures on these agendas. Section 4 specifically attempts to explain these differences by examining the security architectures of these regional organizations.

Following the comparative parts of this study, section five addresses the impact of inter-regionalism on regional commitments to environmental security. The section focuses on the European Union (EU) because, unlike other global powers, such as the United States (US) or China, it is a region rather than a nation-state and the EU has made a commitment to being a global normative power by including commitments to good governance, human rights, the environment, etc. in foreign policy discourse. This case study of EU–Andean Community of Nations (CAN) relations demonstrates that the impacts of inter-regionalism are complex and the existence of different normative views in the field of environmental security can create tensions between regions.

1.1. *Methods and Key Concepts*

This article is based on a review of official policy documents and the websites of 16 regional organizations, in addition to secondary sources. Specifically, research was based on regional policy documents in the field of security and an examination of the mention of environmental security within these documents.

This study raises key concepts in the field of international relations, but to varying degrees and for different purposes. The core concept in this article is 'security' defined as protection from harm. The article differentiates 'hard security' defined as protection from coercion-based threats and 'soft security' defined as protection against non-coercion-based threats. Specifically, the article's primary focus on 'environmental security' defined as protection from environmental dangers, the lack/depletion of strategic resources and conflict over these resources, is generally classified under 'soft security'. 'Human security' broadly defined as the protection of human dignity is a combination of 'hard' and 'soft' security measures. 'Regional policies', another term of primary importance in this article, are defined as those policies that are conceptualized and implemented by regional organizations and they do not include the individual policies of member states. 'Globalization' defined here as the internationalization of political norms is of secondary concern because it merely provides the framework for the article. Similarly, the term 'norms' defined as 'shared political convictions that influence policy discourse/content' is introduced in order to explain expanding security debates and contextualize this study of regional environmental security. Finally, 'development' is referred to as a process that affects our understanding and levels of security within and between polities.

2.   Literature Review: How the Globalization of Security Norms Contrasts Trends in Security Policies

As stated in the introduction above, security paradigms have shifted considerably since the 1980s. Scholars such as Thomas (2001), Newman (2001), and Owen (2004) have documented the historical development of this paradigm since the Brandt Commission's Report opened a debate on the relationship between development and security in 1980 and the United Nations Development Program's (UNDP) 1994 Human Development Report brought the concept of human security to the forefront of international affairs. Specifically, these milestone reports questioned our very understanding of 'security' all the way to its foundations. They focus on three fundamental questions: (1) What is security? (2) For whom is security? and (3) What threatens security?

2.1.   *The Emergence of Human Security as a Conceptual Paradigm*

The immediate answers to these questions have created an inseparable link between security, development, and governance. In response to question one: human security has been defined by the UNDP as 'freedom from fear and freedom from want' (see UNDP, 1994 Human Development Report). As Thomas (2001) argues in her seminal work on human security, this concept entails two facets, one quantitative, and one qualitative. In quantitative terms, 'freedom from want' includes the provision of all material needs for human life, including food, health, education, etc. Conversely, qualitative aspects of human security are related to 'freedom from fear' through the protection of human rights, the protection of physical safety and autonomous control over one's life course. For Thomas,

> . . . human security describes a condition of existence in which basic material needs are met, and in which human dignity, including meaningful participation in the life of the community, can be realized. Such human security is indivisible; it cannot be pursued by or for one group at the expense of another. (p. 162)

Of course, these discourses affect question two: for whom is security? The major shift created by this paradigm is the emphasis on the individual, or people if considered collectively, as the referent(s) for security (see Khong, 2001). No longer do security debates exclusively address territories or states but people-centered policies are promoted.

This of course affected policy responses to question three focusing on threats to security. First, coercive threats to security have been redefined to include non-state actors (organized crime, cartels, terrorism, etc.) and intra-state conflict (civil wars, revolutions, rebellions, etc.) Second, non-coercive threats to security have been incorporated into policy strategies. These include poverty, corruption, discrimination, and of course, environmental issues such as climate change and access to water which are the focus of the next sub-section.

2.2. *Environmental Security: What Place in Security Debates?*

Within the ever-broadening international agenda, environmental security has emerged as pillars of rights-based security norms. Environmental security refers to threats posed by environmental events and conditions to the well-being of individuals, communities, or states (see Barnett, 2011). The development of the early literature on environmental security in the 1990s (see Dabelko & Dabelko, 1995; Homer-Dixon, 1994) often focused on conflicts and violence generated by competition for natural resources. In this regard, scholarship and policies identified and theorized the relationship between environment and security in classical terms. According to Floyd (2008), environmental security emerged as a paradigm because it 'linked to the military'. Floyd specifically identifies the emergence of budget lines for environmental issues in US military spending.

The emergence of environmental security as a global norm is important because it recognizes the role of environmental resources in transnational conflicts. These resources include water (see Conde, 2010), minerals (see Puerta, 2013), and land (see Zoomers, 2010), among others and they have been central elements of recent conflicts in the Middle East (i.e. the Israel–Palestine conflict) and Africa (i.e. the conflicts in Darfur and Congo). In fact, 'environmental security' has shifted the focus in environmental agendas from conservationist discourse aimed at protecting the environment for its inherent moral value, to strategic discussions of how environmental conflicts are inherently human-made and thus, the language utilized in this paradigm reflects the realpolitik approach that has traditionally characterized international security debates (see Zeitoun & Warner, 2006; Koff & Maganda, 2014). Floyd (2008) states that environmental issues have reached the 'equivalent of military problems'.

In terms of these global security debates, for example, threats to environmental security have emerged in terms of global discussions on climate change impacts (see Dabelko, 2009). The recently concluded 21st Conference of Parties to the United Nations Framework Convention on Climate Change (COP21) focused on this issue and global discussions have analyzed the question of responsibilities and potential remedies to climate change impacts within the framework of sustainable development, democratic participation and the promotion of a global green economy. This global discussion recognizes the inter-linkages between development in different parts of the world and interdependencies between regional economies.

In the academic literature, this issue of interdependence has emerged in different ways. Apart from climate change discussions that center on 'ecological shadows' (the resources that a country draws from elsewhere) and 'greenhouse footprints' (emissions generated by a country to sustain its population), the literature has developed beyond classical frameworks related to direct conflict over natural resources. For example, Prins (1990) has argued that environmental issues cannot be addressed through classical policy-making strategies but that concerted policy collaboration between states is necessary to effectively address environmental challenges which are transnational by nature. Dalby (2002) has broadened the notion of environmental security by analyzing 'colonialism' and the 'colonial imagination'. He contends that the world cannot be environmentally separated but that through interconnections, consumption patterns in one part of the world affect the management of strategic resources in another. Moreover, identity formation and the diffusion of economic ideologies affect resource management throughout the world. This even affects how parks and other protected areas are managed in relation to ecotourism or tourism in general. As stated above, however, the main problem with non-traditional security strategies that are norm-based regards effective implementation.

2.3. *Human and Environmental Security and Implementation Difficulties*

The human right to water was passed by the United Nations in 2010. However, this is a right that is very difficult to implement. According to Maganda (2013), water stress remains a vital issue in global affairs that affects about 1.6 billion people, almost one quarter of the world's population. The problem with the implementation of environmental security in general is related to its normative success. On one hand, these norms have emerged because they are inclusive by definition. In exchange, because they address notions of 'justice' they either raise questions related to the need to redistribute access to environmental resources (see Maganda, 2009) or the distribution of costs and benefits (see Shockley, 2011) to address generalized environmental threats.

In this regard, environmental security suffers from the same weaknesses as 'human security'. The aforementioned 'freedom from fear and want' addresses profound issues related to human rights and responsibilities. On one hand, some observers of human security view this approach as a means to support basic human rights (see Ceballos, 2011). Conversely, other scholars suggest that human security offers policy options to states that permit them to address development issues while avoiding recognition of some human rights (see Naranjo Giraldo, 2009). In these cases, basic services such as food and water are viewed through the lens of 'resource management' rather than 'rights protection', thus reducing the security aspects of these policies.

This is a fundamental weakness of human security-based strategies which have not established an international consensus around implementation mechanisms. For example, R2P is a fundamental international norm related to human security which should rest on three principles: (1) conflict prevention, (2) military intervention in the case of attacks on civilian populations, and (3) reconstruction of divided countries. Thus far, R2P has only been sporadically implemented. In part, this has occurred because the norm is vaguely defined. What constitutes a security threat that is so strong that it warrants international responses that disregard state sovereignty? In fact, sovereignty is a key issue in these debates as it is seen as a state's responsibility rather than a right. Having said this, R2P also raises serious implementation issues as noted by scholars such as Spies and Dzimiri (2011). Moreover, so much political debate has focused on military interventions that principles one (conflict prevention) and three (reconstruction) have been virtually ignored.

Similarly, recent scholarship on environmental security has significantly questioned the logic of implementation of securitized environmental policies. For example, Trombetta (2008) has contended that the securitization of climate change mitigation strategies has not necessarily optimized policy implementation. Specifically, Trombetta argues that security frameworks are based on a logic of conflict which does not necessarily apply to environmental issues. For this reason, the securitization of environmental issues is promoting a slow transformation of security logics and provisions due to the introduction of non-confrontational strategies. However, as Floyd (2008) notes, these transformations have been muted by the 'war on terror'.

In fact, nation-state security politics based on hard power have recently been reinforced in global affairs. Since the September 11, 2001 attacks on the US, the 'war on terror' has led to tangible and decisive policies that have included the invasions of Afghanistan and Iraq, broader domestic anti-terror monitoring in the US and the reinforcement of external border controls (see Brunet-Jailly, 2007). The US has also externalized its border security through development programs in the Americas, such as the Plan Sur (Mexico), the Plan Puebla-Panama (PPP), and the Plan Colombia. The PPP, also known as the Mesoamerican Integration and Development Project, was established in 2001. It has included 3.5 billion US dollars of funding in eight development areas including some related to environmental security, such as energy, sustainable development, and disaster prevention and mitigation:

- Energy Sector Integration
- Transportation Integration
- Telecommunications Integration
- Trade Facilitation
- Sustainable Development
- Human Development
- Tourism
- Disaster Prevention and Mitigation

The EU has adopted a similar approach to security in part due to terrorist attacks in the UK, Spain and, most recently, France. Like the US, the EU has focused its attention on coercive security responses through which domestic security issues have significantly affected foreign policy (see Lavanex, 2004). The EU has established a common border control agency (FRONTEX), it has significantly increased inter-state cooperation in anti-crime strategies, and the Schengen Information System has been established with the principle objective of providing information on security threats posed by third country citizens. Like the US, the EU has also externalized its border control policies by implementing conditionality in its trade agreements with other world regions (i.e. the Economic Community of West African States (ECOWAS) and the *Union économique et monétaire ouest-africaine* (UEMOA)) and its development aid (see Koff & Nanga, 2013). A comparison of the EU and US security policies gives credence to what Cornelius, Tsuda, Martin, and Hollifield (2004) have dubbed 'the convergence hypothesis'. These authors argue that state/regional border control policies are increasingly converging because of the emergence of transnational terrorist threats, global demographic trends, similar labor market and welfare state limits, and transnational organized crime. Cornelius et al. (2004) also correctly propose the 'gap hypothesis' which indicates a separation between border security objectives and the results obtained. This gap clearly indicates the limits of coercive nation-state approaches to transnational security.

Consequently, human and environmental security should be viewed in terms of competing globalizations. On one hand, the globalization of security norms has broadened and humanized their objectives but this has led to implementation challenges. Conversely, the globalization of security threats has led to policy retrenchment and the reinforcement of nation-state-based strategies which also face implementation difficulties. How can this paradox be resolved? Recent scholarship on global security has focused its attention on regional security governance. For this reason, this article asks as its main research question: Can regional organizations reconcile the inconsistencies between international security norms and domestic policies with regard to environmental security?

3. Comparing Regional Security Regimes: What Place for Environmental Security?

Regional environmental governance has emerged in recent years as an important paradigm in global environmental politics (see Petit & Maganda, 2012; Weale et al., 2000). This is an important development because environmental issues are generally considered transnational, and not international in nature. While political borders may cause policy implementation problems in Europe, the fact that they reflect past power relationships (North America: notably the Mexico–US divide) and colonial regimes (Africa, Asia, and South America) creates an entirely different level of difficulties with environmental governance in these parts of the world. Specifically, political borders do not respect the geographical boundaries in these continents nor do they reflect demographic or ethnic compositions which has contributed significantly to resource-based conflict in places such as Darfur, Liberia, or Congo.

Global environmental governance, especially at the level of the United Nations, has not been able to suitably address environmental security because this organization is 'international' and it is not well placed to address the 'transnational', defined as localized conflicts/problems that spill over national divides. Some observers of the 2012 Rio+20 Summit have noted this important distinction. For example, the Council on Foreign Relations has reported Dr. Suan Ee Ong's (Nanyang Technological University) claim that the Rio conference erred on the side of breadth while lacking in depth (http://www.cfr.org/world/examining-rio20s-outcome/p28669). The argument here fits the recent trend in security studies presented in Section 2. The practice of global summitry in development and environmental politics reaffirms leaders' commitments to sustainable development norms that are intrinsically linked to human security and environmental security but mechanisms for implementation cannot be identified because consensus-building in these arenas is seemingly impossible. Consequently, the separation between international norms and transnational environmental governance is strikingly evident. This has been shown by Maganda (2009) in her analysis of water security. Maganda compares world maps of socio-economic wealth and access to water and she shows that water access seems dependent on regional distinctions between levels of economic and political consolidation. For this reason, she argues that more attention should be paid to regional water policies. Consequently, this article raises the first research question presented above: Can regional organizations reconcile the inconsistencies between international norms and domestic policies with regard to environmental security?

Much of the broader scholarship on regional environmental governance contends that potential exists for an affirmative response. For example, Ken Conca argues that regional approaches to environmental governance are attractive for four reasons: (1) regions offer hope for political progress where global discussions have stalled, (2) regions offer superior conditions for common property resource management, (3) regions are more conducive to promoting norm diffusion, and (4) regional approaches may be a foundation for a cumulative approach to building global environmental governance (Conca, 2012). Alternatively, Debarbieux (2012) indicated that 'knowledge regions' can promote sustainable development through the establishment of spaces of dialogue and even common policy identities.

In fact, regional environmental agendas are quite developed throughout the world. Table 1, which is based on a review of policy documents from the regional organizations included in the table, shows the different policy objectives of environmental regimes in 16 regional

Table 1. A comparison of 16 regional environmental and security agendas

Regional organization	Regional environmental agendas	Regional security agendas (environmental aspects in italics)
African Union (AU)	Biodiversity, climate change, energy, environmental	Small arms, weapons of mass destruction, counter-terrorism, conflict early warning system, landmines, cross-border security, post-conflict reconstruction and peace-building (*including the promotion of sustainable development*)
Common Market for Eastern and Southern Africa (COMESA)	Climate change, food security	Conflict early warning system, war economies approach, human security and good governance (transparency and promotion of democracy)
East African Community (EAC)	Climate change, bio safety, water management, minerals and mining	Cross-border crime, drug trafficking, small arms, counter-terrorism and anti-piracy, research on resource-based conflict
Economic Community of Central African States (ECCAS)	Forest conservation, common energy policy, fishing	Conflict early warning system, election monitoring
Economic Community of West African States (ECOWAS)	Water resources, energy	Small arms, peacekeeping, border security
Southern Africa Development Community (SADC)	Food security, natural resource management, sustainable development, disaster prevention	Protect against instability and intra- and inter-state conflict and aggression, conflict early warning system, intelligence cooperation, peacekeeping and peace-making, cross-border crime, conflict prevention, protection of human rights, migration governance, *disaster prevention*
Mercado del Sur (MERCOSUR)	No regional policy arenas recognized	Democratization and military–civic relationships; inter- and intra-state conflict; terrorism, organized crime, drug trafficking
CAN	Climate change, food security, water, disaster prevention, biodiversity	Common Andean Policy on External Security; establishment of a Peace Zone in the Andean Community; limiting military spending in order to use those funds for social investment purposes; intensifying cooperation to fight terrorism, illegal arms trafficking and drug trafficking; *harmonization of security policies with social development, environmental and biodiversity management, and human rights*
Union of South American Nations (UNASUR)	*Energy security*	Defense cooperation, trafficking and organized crime, democratic stability, economic security, *energy security*

(Continued)

Table 1. Continued

Regional organization	Regional environmental agendas	Regional security agendas (environmental aspects in italics)
Sistema de la integración centroamericana (SICA)	Water (including climate change impacts), food security, sustainable energy	Democratic security including rule of law and transparency, strengthening the role of civil society in security policy-making, anti-corruption, anti-terrorism, small arms, drug trafficking, *disaster mitigation and prevention*
North American Free Trade Agreement (NAFTA)	Cross-border water management, climate change, pollutants, protection of ecosystems	Migration, terrorism, drug trafficking, organized crime
Organization of American States (OAS)	Biodiversity, energy, climate change, water resource management	Drug trafficking, human trafficking, terrorism, cyber security, disarmament, landmines, conflict resolution (including Peace Fund)
Caribbean Community (CARICOM)	Water resources management, sustainable land management and integrated watershed and coastal areas management waste management: solid, liquid, hazardous, biomedical and electronic waste sustainable consumption and production, eco-efficiency and renewable energy environmental and social impact assessments climate change	Organized crime, small arms, drug trafficking, financial crime, cyber-crime, corruption, human trafficking and smuggling, *disaster prevention*, pandemics, *climate change*, irregular migration
Association of Southeast Asian Nations (ASEAN)	Climate change, haze pollution, water resources management, disaster prevention	Comprehensive security including political development, peace through norm diffusion, conflict prevention, conflict resolution, post-conflict peace-building through common actions such as humanitarian assistance
European Union (EU)	Biodiversity, water governance, soil and land management, waste management, climate change, air pollution, natural resource management, sustainable production and consumption, sustainable energy, policy coherence	Counter-terrorism, organized crime, human trafficking, sexual exploitation, irregular migration, external peace-building, a responsible neighbor, human rights, *climate change*, development cooperation, disaster relief and humanitarian aid, trade
Arctic Council	Climate change, biodiversity, oceans	*Climate change*, strategic resources including oil, minerals and natural gas, governance of shipping lanes, territorial disputes

Source: Table compiled by author.

organizations. This table illustrates the developed state of regional environmental governance as most regional organizations have made significant commitments to sustainability in key environmental areas such as the protection of biodiversity, water resource management, climate change mitigation, and energy policies. This seems to reinforce the notion that regions are well placed to address environmental issues. On one hand, their transnational nature offers opportunities for collaboration in innovative management schemes. For example, the EU has implemented a Water Framework Directive that has facilitated the governance of cross-border water basins. Similarly, the Association of Southeast Asian Nations has made smog and air pollution a priority on its agenda because it is a transnational phenomenon that affects various states in the region.

While Table 1 does recognize regional commitments to sustainable development, it also highlights a broader problem related to environmental security. Section 2 has shown that this norm has been accepted internationally and the table also demonstrates that regional organizations have made important commitments to this norm through the recognition of environmental security in key regional development programs. Nonetheless, the third column in the table lists the key issues in regional security regimes. Environmental issues are highlighted in italics in this column. It is clear that 'regional environmental security' has not yet been fully integrated into general regional security strategies. For this reason, it seems evident that world regions have recognized environmental issues as political priorities and potential exists for them to address globalized security debates. However, these regions have not yet operationalized 'environmental security' within their regional security agendas.

Thus, one must ask: How can we discuss 'environmental security' separately from broader security debates? The policy arenas listed in column three of the table seem to reinforce the conclusion presented in Section 2: the globalization of security threats has narrowed security policies which largely focus on inter- and intra-state conflict, organized crime, trafficking, and terrorism. The AU has vaguely included sustainable development as part of its peace-building post-conflict reconstruction program and the Andean Community has impressively linked sustainable development to a broader security agenda that includes human rights, anti-poverty strategies, and the rights of indigenous populations. Largely, however, regional environmental security policies are limited to disaster prevention in developing regions (SADC, SICA, CARICOM) and climate change policies in wealthier and more consolidated regions (EU and Arctic Council). UNASUR focuses on energy following the interests of member states such as Venezuela and Ecuador which are exporters of oil, Bolivia which exports gas, and regional hegemon Brazil which exports oil and alternative fuels such as palm oil.

For this reason, it seems that regions presently hold limited potential for the implementation of environmental security norms. The trends illustrated by Table 1 indicate that regional security agendas, both general and environmental, seem to be driven by the interests of member states rather than international norms. Consequently, regions should be viewed as extensions of their members' security personalities and their security agendas seem to reflect the accumulated interests of their member states. In short, they seem to be extensions of state-based security policies rather than autonomous polities with their own security agendas based on globalized norms which inform their member states. This leads to the second research question listed above: Why is environmental security not prominent in regional security agendas?

4. Comparing Regional Security Architectures: What Opportunities for Implementation of Environmental Security Strategies?

The previous section has shown that environmental security is marginalized in most regional security agendas. This suggests that nation-states remain dominant actors in transnational security debates. An emerging literature on regional security regimes has documented this situation in great detail. Fawcett (2006) has noted that 'security regionalisms' have become prominent in specific security fields (i.e. border controls, migration policies, anti-terrorism, etc.) but that, in general, these organizations have not developed strong regional security institutions. Similarly, van Langenhove et al. (2009) have recognized the significant contributions that numerous regional organizations have made to the protection of human security in different world regions. They too, however, have contended that the successful implementation of human security policies depends on multilevel governance and coordinated actions at the supranational, national, and sub-national levels.

These studies, among others, provide clues as to why environmental security has not figured prominently in regional security regimes. They have indicated that unlike other policy arenas, such as environmental management, where regions have taken leading roles in policy implementation, security has remained an issue dominated by member states which are hesitant to cede authority to regional organizations over hard security questions. This is evident in the EU (which according to Kirchner and Dominguez has the most developed regional security architecture in the world) where Common Foreign and Security Policy (CFSP) has been one of the policy arenas in which supranationalism has been slowest to develop (see Hix & Hoyland, 2011).

For this reason, it seems that security architectures influence regional security strategies more than international norms such as environmental security. Even the EU, which has attempted to carve a niche for itself in global affairs through the operationalization of commitments to global norms such as human and environmental security, is characterized by CFSP that is driven by the interests of member states. Smith (2004), among others, has analyzed the establishment of the EU's CFSP through the lens of multilevel governance. He contends that CFSP results from the combination of domestic politics of member states and institutional mechanisms. Rarely does this scholarship mention the evolution of global norms in the formulation of CFSP.

The most complete comparative study in this field has been conducted by Kirchner and Dominguez (2011). These authors examined the relationship between regional security performance and the domestic economic and political conditions of 14 regional organizations' member states. In general, the authors of this study noted that member states have readily supported regional organizations in the definition of regional security policies. However, their support has waned through the policy coordination and implementation stages. Consequently, member states supported regional security discourse without necessarily supporting operationalized regional security policies.

Also, Kirchner and Dominguez note that significant variance exists between the definition of security strategies in different policy arenas and the depth of implementation of these strategies. Of the 14 regional organizations included in the study, nine were ranked with low levels of regional security governance. Only the EU scored highest in both the range of policy arenas covered in the regional security regime and the ability to implement these policies (including coordination and implementation mechanisms). Mixed regimes, such as UNASUR, CAN, and CARICOM, had well-defined security regimes but their commitments to implementation stopped at the policy coordination stage (see Kirchner & Dominguez, 2011).

Of course, these findings indicate that regional security agendas remain dominated by organizations' member states. The present article confirms that regional security architectures vary significantly and this affects both the breadth and depth of these security regimes. It is difficult for environmental security policies to emerge in agendas that are largely dominated by state actors following domestic incentives and conflict-based security logics.

Table 2, which is based on an examination of institutional frameworks of 16 regional organizations, presents the security architectures for these organisms. The analysis is based on the research conducted by Kirchner and Dominguez. It classifies the case regional organizations into three groups: (1) 'open security structures' that are defined by regional authorities and characterized by direct public participation. This legitimizes their activities and confirms that regional actors

Table 2. Comparison of 16 regional security architectures

Regional organization	Regional security architecture
Open security structures—public participation	
AU	Peace and Security Council: members represent Africa's Sub-regions; African Peace and Security Architecture: parliamentary and civil society inclusion
COMESA	Committee on Peace and Security; Ministries of Foreign Affairs; Programme on Peace and Security: Civil Society Participation, Inter-Parliamentary Forum, Committee of Elders
EU	External Action Service and Justice and Home Affairs
Weak regional security structures—lack of public participation	
EAC	Sectoral Council on Interstate Security: member states
ECOWAS	Mediation and Security Council; ECOWAS Conflict Prevention Framework
SADC	Organ for Politics, Defense and Security which includes member states' ministers of defense
CAN	High-Level Group on Security and Confidence-Building
UNASUR	Energy Council of South America
SICA	Central American Security Commission (autonomous)
Ad hoc bilateral negotiations or formal/informal consensus-building	
NAFTA	Ad hoc bilateral relations
OAS	Secretariat for Multidimensional Security, Security consensus through formal and informal structures
CARICOM	CARICOM's Implementation Agency for Crime and Security; consensus through formal and informal institutions
MERCOSUR	Security consensus through formal and informal structures
ASEAN	Security consensus through formal and informal structures
ECCAS	Peace and Security Council; ad hoc bilateral relations
Arctic Council	Security consensus through formal and informal structures

Source: Table compiled by author based on policy documents.

are defining policies and not simply implementing those strategies forwarded by member states, (2) 'weak security structures' in which regional institutions exist but they are dominated by member state actors, and (3) organizations in which security structures are lacking and regional policy is made through ad hoc bilateral negotiations or formal/informal consensus-building.

This table confirms many of the conclusions presented by Kirchner and Dominguez. Seven of the organizations examined here are characterized by ad hoc bilateral negotiations between member states or consensus decision-making through formal or informal channels (ECCAS, MERCOSUR, NAFTA, OAS, CARICOM, ASEAN, Arctic Council). Six more are characterized by relatively weak institutional structures (EAC, ECOWAS, SADC, CAN, UNASUR, SICA) that loosely coordinate security strategies in regional organizations and only three regions have institutionalized and open security structures that promote public participation and legitimate decision-making (AU, COMESA, EU).

Because regional organizations, especially those in the developing world have weaker security architectures, it is not surprising that their strategies are more closely tied to the interests of member states than international security norms. A more systematic comparison of these regimes' characteristics illustrates this point. In order to understand the place of environmental security in regional security agendas, this article utilizes scales to compare this (dependent) variable to regional environmental policies and regional security architectures. 'Regional environmental policies' have been measured in terms of their breadth by counting the issue arenas that regional organizations have included in their environmental policies. This approach highlights the general value of 'environmental security' present within a given regional organization. Conversely, 'environmental aspects of security strategies' and 'regional security architectures' have been operationalized by utilizing the model proposed by Kirchner and Dominguez which examines policy definition, policy harmonization, and policy coordination. The article proposes an ordinal scale in which:

0 = no regional definition of security
1 = regional definition but no policy harmonization or coordination
2 = regional definition and harmonization but no policy coordination
3 = regional definition, harmonization and coordination

Figure 1 compares these variables across the 16 regional organizations presented in Tables 1 and 2. The lines in the figure illustrate a weak but present relationship between the breadth of regional environmental agendas and the place of environmental issues in regional security strategies. However, the trends illustrated by this figure illustrate a much closer relationship between the character of regional security architectures and the importance of environmental issues in regional security agendas. Given that 13 of the 16 regional security architectures examined in this article are considered 'weak' at best, this would seemingly reinforce the argument that the 'globalization of security norms' is weak and it would seem to explain the difficulty with promoting environmental security more prominently in regional security debates. This point is further illustrated below through a qualitative analysis of the EU and its inter-regional policies.

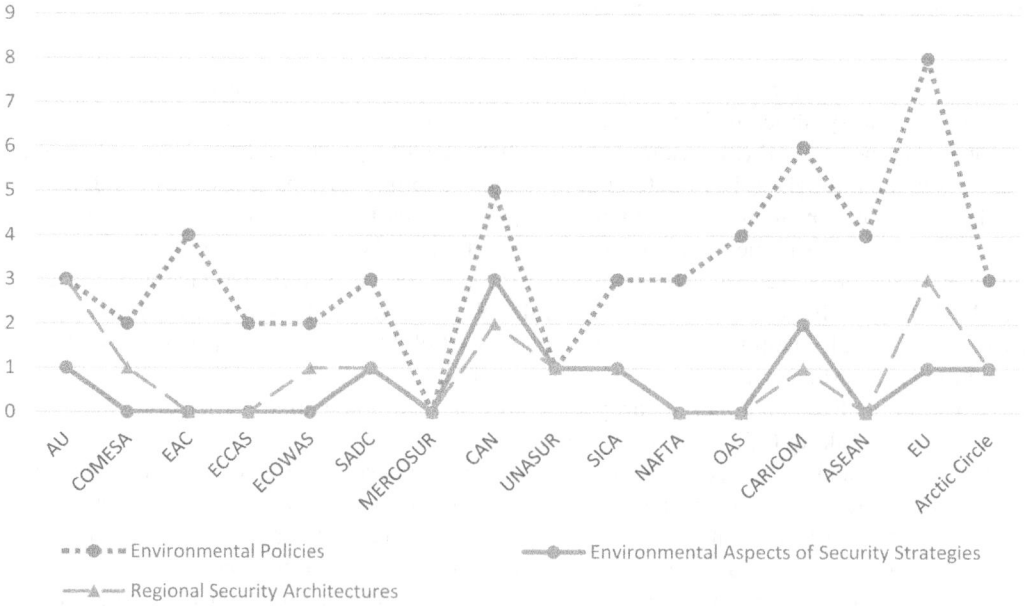

Figure 1. A comparison of regional environmental security agendas with regional security architectures.

5. Inter-regionalism and Environmental Security: A Discussion of EU–CAN Relations

This section of the article asks: what is the impact of inter-regionalism on regional environmental security? Specifically, it inquires whether inter-regionalism promotes the globalized norms addressed in the literature review or whether regions behave like macro-states, reinforcing hard security approaches. In order to complement the comparative analysis presented above, this section presents a case study of the EU. The EU represents a critical case in the context of this study. On one hand, the EU has positioned itself as a 'normative power' in the world (see Pace, 2007) through open diffusion of 'core values' such as human rights and democratic governance. In the field of environmental governance, the EU has criticized other world powers such as the US and China for their relative inaction in the field of climate change mitigation. Moreover, as Kirchner and Domiguez have shown, the EU has the most complete regional security architecture in the world in terms of regional definition and regional implementation. For these reasons, the EU should present an example of 'best practice' in terms of implementing environmental security within its regional security agenda.

The article discusses EU cooperation with the CAN. This relationship has been chosen for two reasons. First, it is an example of symmetrical inter-regionalism (see Roloff, 2008) because it involves cooperation between two regional organizations. Second, as Table 1 illustrates, the Andean Community has internally demonstrated a commitment to environmental security on two levels. First, it has a well-developed environmental agenda with policies on topics such as climate change, food security, water, disaster prevention, and biodiversity. Second, the CAN has integrated environmental dimensions in its security agenda and it actively attempts to harmonize security policies with social development, environmental and biodiversity management, and human rights. These characteristics are recognized in the scholarship on CAN regional integration by scholars such as Marquez and Romero (2003) and König (2013). For these structural reasons, one would expect environmental security to emerge in EU–CAN cooperation, also within the framework of security discussions.

A review of EU inter-regional agreements (based on official policy documents and announcements posted on the EU website and those of its regional partners) indicates a more complex situation. It is true that the EU has promoted sustainable development and environmental conservation through its development strategies, notably, the Lomé III and Lomé IV Conventions and the Cotonou Agreements. The EU has included food security and water security in its development programs and it has demonstrated a strong commitment to the Millennium Development Goals in both its development and trade relations with ACP (Africa, Caribbean, Pacific) countries.

Nonetheless, even though the EU has promoted environmental security in development and trade, environmental concerns have not emerged in EU security agendas nor in EU partnership negotiations with other world regions. For example, the Barcelona process (1995–2005) which aimed to establish a free trade zone in the Mediterranean Basin operationalized the externalization of borders and migration controls because all non-EU Mediterranean states participating in the initiative committed to support the EU's actions to prevent clandestine migration. Since 2005, when the Euro-Mediterranean Union was formed, the language of this approach has been softened as the union discusses 'the promotion of regular migration'. Nonetheless, the implementation of these initiatives has not changed as the EU and member states, such as France and Spain, have invested heavily in anti-immigration and anti-trafficking measures in third countries. Conversely, Marín Durán and Morgera (2012) have noted that even though environmental norms are part of the discourse of the Euro-Mediterranean Union, 'Euro-

Mediterranean Associations do not establish special procedures for the implementation of environmental commitments, nor for resolving environment-related disputes' (p. 81). In this regard, the EU has adopted normative discourse but it has not operationalized it through implementation mechanisms. Instead, it has reinforced member state security interests.

The EU's relationship with the Andean Community of Nations is more complex, in part due to the CAN's commitment to environmental security. A cooperation agreement signed in 1993 included numerous policy arenas related to environmental security. In general, the agreement identifies the promotion of sustainable development as a main objective of EU–CAN cooperation. Moreover, specific goals were enumerated in relation to individual policy fields. For example, 'protection and improvement of the environment' and 'rational use of natural resources' were included in the agreement under the auspices of 'cooperation in science and technology'. The agreement also included specific focus on mining, energy, forestry, and fisheries.

In the EU's Andean regional action plan from 2003 to 2006, environmental security also figured prominently. Specific budgets (amounting to nine million euros) were allocated to disaster prevention and disaster relief. Other identified priorities included: climate change, water, biodiversity, and forests. Moreover, the agreement stipulated that environmental impact assessments would be carried out when appropriate at the project level.

Despite this seemingly solid collaboration in environmental arenas, tensions have arisen in the EU–CAN relationship due to different positions on important aspects of environmental security. One of the most significant issues which has actually blocked deeper EU–CAN cooperation is water privatization. The EU has pushed for water privatization in its cooperation agreement negotiations with the CAN. In 2007, Bolivian President Eva Morales prominently withdrew from EU–CAN trade negotiations because they included water privatization.

This dispute manifested itself on a much larger scale in 2010. As Carmen Maganda's contribution to this special issue has shown, the EU has not adequately included water security in its foreign policies. Maganda's article documents, how EU member states unanimously abstained from the UN vote on the human right to water in 2010 which was heavily supported by the Andean Community following its introduction to the United Nations by Bolivia.

In addition to its lack of support for water security, the EU has been criticized for focusing almost its entire cooperative security agenda with the CAN on the fight against drug trafficking. Smaller initiatives focus on combatting illegal immigration, the trafficking of small arms and terrorism. In all these fields, EU rhetoric claims to support policy strategies that present

> a balanced, multilateral, inclusive and non-selective approach to this issue. In this vein Ministers stressed the importance of stepping up cooperation and of strengthening efforts to address, in a comprehensive way, all relevant supply and demand issues, including political and social stability, security and sustainable development. (http://www.eu2005.lu/en/actualites/conseil/2005/05/26ueand-cc/index.html)

However, when the CAN attempted to introduce rights-based approaches to these issues into the most recent cooperation agreement negotiations, the EU supposedly refused (http://www.comunidadandina.org/ingles/exterior/migrations.htm). This has created discontent among some members of the Andean Community of Nations, especially among NGOs that focus on the protection of biodiversity and water management. Curiously, 'the environment' seems to have been demoted in EU–CAN cooperation as it now has been incorporated under to the more general theme 'social and economic cohesion' in the EU's 2007–2013 regional strategy paper. The development seems to indicate that the tension over water management and privatization issues seems to have affected the EU–CAN relationship in the field of environmental security. Moreover, Andean interest groups and non-governmental organizations have mobilized over these issues. In response, the EU and the CAN have established a civil society forum in which participants discuss varying aspects of EU–CAN cooperation, including environmental security. Such measures have the potential to help reconcile the promotion of globalized security norms as described above, with coercion-based security strategies presently adopted by most regions. This is discussed in the conclusion below.

6. Conclusion

This article has contended that the world is witnessing competing globalizations in security affairs. On one hand, security norms have globalized beyond nation-states as human security and environmental security have emerged as powerful concepts in the international arena. Conversely, the globalization of security threats has reinforced nation-state security strategies, at times to the detriment of international norms.

In line with the emerging literature on regional security governance, this article supports the premise that regional security strategies are well placed to promote new security norms while simultaneously addressing transnational security threats because these organizations are supranational in nature but they are characterized by transnational governance mechanisms (i.e. multi-level governance). The problem with many of these regimes, however, is related to their weak regional governance architectures. First, regional security regimes are characterized by weak institutions which permits member states to reinforce their own security preferences. This is dangerous because national security policies are not only maintained, but they are strengthened since they have been politically legitimized through approval by regional bodies. This point is clear in the presentation of the EU's security agenda in which inter-regional agreements with Mediterranean and African regional organizations aim to reinforce security more than norm-based development (see Miranda, Pirozzi, & Schäfer, 2012).

Unfortunately, nation-state-based strategies do not satisfactorily address transnational threats as recent events in West Africa (civil strife in Mali), East Africa (terrorism in Kenya and piracy in Somalia), and Europe (Paris terrorist attacks) have shown. Moreover, the lack of focus on social cohesion within regions, means that short-sighted security policies do not address the underlying political, economic, and environmental vulnerability that causes security threats.

This leads to the second weakness of these regional security architectures: they have not properly addressed the issue of policy coherence. The sections above illustrate that environmental security has not been integrated into more general security regimes. Environmental concerns significantly affect the state of security in any given region. Unfortunately, because these issues are not linked to security policies in terms of definition, coordination, or implementation, sustainable development strategies are often undermined by security practices. The management of cross-border water resources presents an excellent example of this situation. When borders are closed by national and supranational leaders due to the implementation of anti-trafficking or migration control policies, how can local populations be expected to coordinate and implement effective trans-boundary water management policies? How can adequate infrastructure be constructed? Security policies aimed at protecting national citizens are making illegal trafficking more profitable and environmental security more expensive and difficult to achieve.

The comparative analysis presented above, in combination with the case study of EU–CAN inter-regionalism, does, however, offer some clues for potential paths to the establishment of regional environmental security. Section 3 of this article has indicated that the relative exclusion of environmental security policies from regional security agendas can be explained by the weakness of regional security architectures. The central question that needs to be addressed is: 'how can these regional security architectures be reinforced in such a way that they construct their own identity and no longer simply reflect the interests of member states?'

The EU–Andean Community inter-regional relationship would suggest that regions can successfully promote environmental security when legitimized policy strategies are enacted. The EU and the CAN both have demonstrated a normative commitment to environmental security in their internal politics. Unfortunately, this commitment within their collaborative relationship seems to have been overshadowed by tensions generated by trade and hard security considerations. Nonetheless, a recent development may provide a way forward. The EU–Andean Community Civil Society Forum was established in 2005 in order to allow for an open and transparent exchange of views on EU–Andean relations. In April 2007, before the opening of negotiations for an Association Agreement, civil society organizations were invited to participate in an assembly in order to discuss negotiations. The engagement has been maintained throughout the negotiation process.

The EU–CAN civil society forum, like similar initiatives, can be considered a starting point for such democratic input that, if successful, could be implemented in other world regions. In order for world regions to strengthen their operationalized commitments to environmental security, measures should be taken to reinforce their overall security architectures. This can be accomplished through public participation in policy-making and citizen oversight in policy implementation and evaluation processes. Thus far, virtually all of the regional organizations examined in this study suffer from a democratic deficit. As long as citizens are not informed of regional security policies and how they can affect them, nation-states will continue to dominate regional regimes, especially in the field of security which has generally been characterized by realist perspectives based on national interests. Regional security architectures cannot be strengthened without the evolution of regional security identities. The bases for these identities can only be the emergence of regional citizenship identities. People-based security norms, such as human security and environmental security, can only be implemented when people are involved in decision-making processes. Unless citizens call for and join more participative regional security processes, there are few incentives for nation-states to cede authority to regional regimes. Regional organizations can reconcile competing globalizations in the field of security. However, they need to first operationalize this potential through more inclusive governance structures.

Disclosure Statement

No potential conflict of interest was reported by the author.

References

Barnett, J. (2011). *The meaning of environmental security*. London: Zed Books.
Brunet-Jailly, E. (2007). *Borderlands*. Ottawa: University of Ottawa Press.
Ceballos Medina, M. (2011). La politica migratoria de Ecuador hacia Colombia: Entre la integración y la "contención". *Regions & Cohesion, 1*(2), 45–77.
Conca, K. (2012). The rise of the region in global environmental politics. *Global Environmental Politics, 12*(3), 127–133.

Conde, G. (2010). Reconfiguraciones políticas y territoriales del conflicto entre Israel y Palestina. In C. Puerta & J. C. Velez Rendon (Eds.), *Frontera y Reconfiguraciones Regionales: RISC 2009* (pp. 107–136). Brussels: PIE-Peter Lang.

Cornelius, W., Tsuda, T., Martin, P., & Hollifield, J. (Eds.). (2004). *Controlling immigration*. Palo Alto, CA: Stanford University Press.

Dabelko, G. (2009). Planning for climate change: The security community's precautionary principle. *Climatic Change, 96*, 13–21.

Dabelko, G., & Dabelko, D. (1995). Environmental security: Issues of conflict and redefinition. *WW Environmental Change and Security Project Report, 2*(1), 3–13.

Dalby, S. (2002). *Environmental security*. Minneapolis: University of Minnesota Press.

Debarbieux, B. (2012). How regional is regional environmental governance? *Global Environmental Politics, 12*(3), 119–126.

European Commission. (2007). *Andean community regional strategy paper 2007–2013*. E/2007/678. Brussels: Author.

Fawcett, L. (2006). *Regional governance architecture and security policy*. FES Briefing Paper. Berlin: Friederich Ebert Stiftung.

Fawcett, L. (2013). *Security regionalisms: Lessons from around the world* (RSCAS Working Paper 2013/62). Fiesole: European University Institute.

Floyd, R. (2008). The environmental security debate and its significance for climate change. *The International Spectator, 43*(3), 51–65.

Hix, S., & Hoyland, B. (2011). *The political system of the European Union* (3rd ed.). New York, NY: Palgrave Macmillan.

Homer-Dixon, T. (1994) Environmental scarcities and violent conflict: Evidence from cases. *International Security, 19*(1), 5–40.

Khong, Y. F. (2001). Human security: A shotgun approach to alleviating human misery? *Global Governance, 7*(3), 231–236.

Kirchner, E., & Dominguez, R. (2011). *The security governance of regional organizations*. London: Routledge.

Koff, H., & Maganda, C. (2014). Water security in cross-border regions: What relevance for regional human security regimes? In D. Garrick & G. Anderson (Eds.), *Water security and federal rivers* (pp. 325–338). Cheltenham, UK: Edward Elgar.

Koff, H., & Nanga, E. (2013, November 14–15). *Policy coherence for development for whom?: Examining the development-migration nexus in regional security regimes*. Paper presented at Nordic Conference for Development Research, Helsinki, Finland.

König, C. (2013). *The environment in the Andean community and MERCOSUR* (UNU CRIS Working Paper 2013/3). Bruges: UNU CRIS.

van Langenhove, L., Vigilante, A., Fanta, E., Felício, T., Ferro, M., Scaramagli, T., & Tavares, R. (2009). *The regional dimension of human security: Lessons from the European Union and other regional organizations* (Garnet Policy Brief 9).

Lavanex, S. (2004). EU external governance in 'wider Europe'. *Journal of European Public Policy, 11*(4), 680–700.

Maganda, C. (2009). Regiones, recursos y responsabilidades: reflexiones desde la ética ambienta frente a la problematica ambiental global. In C. Puerta & J. C. Velez Rendon (Eds.), *Frontera y reconfiguraciones regionales: RISC 2009* (pp. 89–106). Brussels: PIE-Peter Lang.

Marín Durán, G., & Morgera, E. (2012). *Environmental integration in the EU's external relations: Beyond multilateral dimensions*. Oxford: Hart.

Marquez, G., & Romero, L. (2003). Ecología, Ambiente y Relaciones Colombo-Venezolanas. In S. Ramírez & J. M. Cadenas (Eds.), *La Vecindad Colombo-Venezolana* (pp. 329–344). Bogotà: Universidad nacional de Colombia.

Miranda, V., Pirozzi, N., & Schäfer, K. (2012). *Towards a stronger Africa-EU cooperation on peace and security: The role of African regional organizations and civil society* (IAI Working Paper 12/28). Rome: Istituto Affari Internazionali.

Naranjo Giraldo, G. (2009, November 3–5). *Políticas migratorias y condiciones de seguridad humana de la población desplazada, refugiada y migrantes irregulares en la frontera Colombia-Venezuela*. Paper presented at 2010 Conference of the Consortium RISC, Medellín, Colombia.

Newman, E. (2001). Human security and constructivism. *International Studies Perspectives, 2*, 239–251.

Owen, T. (2004). Proposal for a threshold-based definition human security—conflict, critique and consensus: Colloquium remarks and a proposal for a threshold-based definition. *Security Dialogue, 35*, 373–387.

Pace, M. (2007). The construction of EU normative power. *JCMS: Journal of Common Market Studies*, *45*(5), 1041–1064.

Petit, O., & Maganda, C. (2012). *Strategic natural resources governance: Contemporary environmental perspectives*. Brussels: PIE-Peter Lang.

Prins, G. (1990). Politics and the environment. *International Affairs (Royal Institute of International Affairs 1944-)*, *66*(4), 711–730.

Puerta Silva, C. (2013). *Stratégies et politiques de reconnaissance et d'identité: Les Indiens wayuu et le projet minier du Cerrejón en Colombie*. Brussels: PIE-Peter Lang.

Roloff, R. (2008). In interregionalism in theoretical perspective: State of the art. In J. Rüland, H. Hänggi, & R. Roloff (Eds.), *Interregionalism and international relations: A stepping stone to global governance?* (pp. 17–30). London: Routledge.

Shockley, K. (2011). Divergent principles, development rights, and individualism in the Greenhouse Development Rights framework. *Regions & Cohesion*, *2*(1), 1–24.

Smith, M. (2004). Toward a theory of EU foreign policy-making: Multi-level governance, domestic politics, and national adaptation to Europe's common foreign and security policy. *Journal of European Public Policy*, *11*(4), 740–758.

Spies, Y., & Dzimiri, P. (2011). A conceptual safari: Africa and R2P. *Regions & Cohesion*, *1*(1), 32–53.

Thomas, C. (2001). Global governance, development and human security: Exploring the links. *Third World Quarterly*, *22*(2), 159–175.

Trombetta, M. J. (2008). Environmental security and climate change: Analysing the discourse. *Cambridge Review of International Affairs*, *21*(4), 585–602.

United Nations Human Development Report. (1994). *Redefining security*. New York, NY: United Nations Secretariat.

Weale, A., Pridham, G., Cini, M., Konstadakopulos, D., Porter, M., & Flynn, B. (2000). *Environmental governance in Europe: An ever closer ecological union?* Oxford: Oxford University Press.

Zeitoun, M., & Warner, J. (2006). Hydro-hegemony—a framework for analysis of trans-boundary conflicts. *Water Policy*, *8*, 435–460.

Zoomers, A. (2010). Globalisation and the foreignisation of space: Seven processes driving the current global land grab. *Journal of Peasant Studies*, *37*(2), 429–447.

Water Security Debates in 'Safe' Water Security Frameworks: Moving Beyond the Limits of Scarcity

CARMEN MAGANDA

ABSTRACT *What relevance is there to discuss water security issues in cases where water availability and accessibility do not seem to be a problem? This is the main question that guides this article which searches for answers from official core water debates in upstream or 'safe' water access (or 'water-rich') countries. If a norm such as water security or the human right to water and sanitation is to be universally accepted, then it needs to be adopted by wealthy and powerful countries/regions that are water-rich, and it should guide policies in both their domestic and foreign policies. If these polities do not support this norm, then operationalization becomes a serious challenge and water security debates will keep reflecting power imbalances in global affairs. Empirically, this article examines current water management strategies in the European Union with specific focus on the Grand Duchy of Luxembourg.*

1. Introduction

Water security has generally been associated with water scarcity in global environmental debates. It has also been discussed within the framework of development and the global fight against poverty. If we follow this 'water security' approach defined as a concept unconditionally tied to water scarcity, then we may exclusively look at specific developing regional cases that are characterized by such scarcity. Instead, this article contends that water security should matter to all political actors, without distinction between levels of local water resources because it is emerging as an important norm in international discussions linked to human rights and human security.

Therefore, the article argues that in order to better understand water security today, we should also analyze what is being discussed in security agendas in those countries with 'safe' water access, also called 'water-rich' states. Considering that not all the nations that are upstream or those that have plentiful water supplies do necessarily have to be 'safe water access countries', due to other problems related to security that may not refer exclusively to quantity of available water, such as serious water-quality issues or extreme natural phenomena tied to climate change effects (the so-called disasters). For example, it is complicated to characterize the Netherlands as a 'water-safe country', despite the fact that it is water-rich, given the nation's extreme vulnerability to flooding.

It is also important to consider the effects of time and change in this analysis of security because a nation that seems 'water-safe' today may be vulnerable tomorrow to major transformations due to climate change, population growth, urbanization, economic development, or other forces.

This article discusses water security in countries where this norm is seemingly not needed. It builds on some highly relevant questions such as: What does water security (and insecurity) mean in industrialized/water-rich countries today? How do domestic or regional positions on water security affect foreign policy? The specific research question to which this article responds asks: What value do water security norms have if they are not universally practiced, especially in industrialized water-rich countries that wield power in global affairs?

1.1. *Research Design and Methods*

This research focuses on a case study: the Grand Duchy of Luxembourg. This case can be considered critical not only because water supplies seem to meet existing demands and needs, even though water quality is an issue (Maganda, 2013), but also because it is one of the smallest states in Europe characterized by high economic and demographic growth rates.[1] Using data from interviews with border water and environmental officials in Luxembourg, ethnographic literature and reports from the European Union (EU), I aim to build a theoretical and empirical reflection on water officials' perceptions and behavior on water security issues. Special focus will be given to national positions on water security related to transboundary waters and regional agendas. This will be compared to national positions on water security in development cooperation programs abroad. The article states that water security is a broader approach than water governance[2] which should be universally supported, even in those countries with more 'safe' water access. Moreover, I propose an expanded definition of 'water security' that moves beyond fixed indicators to encompass a broader notion ultimately grounded in the human right to water.

The article is divided into four sections. Following this introduction, Section 2 will present the key terms and conceptual framework related to water security and the need to go beyond water scarcity. It discusses the role of norms in international relations. Section 3 then explores the water security agenda of the EU. Specifically, this part examines the region's commitment to the concept of water security. Section 4 will refer to the case study: a political ethnography of Luxembourg's water security discussions in national and foreign affairs. Finally, concluding thoughts will be presented in Section 5.

2. Overview of Water Security in Poor and Rich Countries. The Need To Go Beyond Water Scarcity

In many public debates, water scarcity and water insecurity are often utilized interchangeably. It is important to make the distinction between these two different concepts. The first one can be defined as a shortage in the availability of fresh water (or the lack of the necessary infrastructure to take water from rivers and aquifers), relative to demand which can directly affect access to safe drinking water, public health, food security, economic and environmental well-being conditions (Taylor, 2009). But water scarcity is also linked to slow economic development with potential to promote civil discontent. Therefore, many scholars and international organizations work hard to define measures of water scarcity to inform water policy and hopefully help allocate resources to mitigate the cited effects.

Instead, 'water security' is a more normative concept which is fairly new, with multiple interpretations that cannot be easily operationalized. For example, the relationship between water security and elements related to both vulnerability (Scott et al., 2012) and human rights issues remains opaque (Gerlak & Wilder, 2012). This term was coined in the beginning of the century as the result of the intense debates surrounding the critical access to safe water and sanitation for everyone. In 2000, the Global Water Partnership (GWP) produced a key document entitled, 'Towards water security: A framework for action', which was presented as a principle text in the 2nd World Water Forum in The Hague. This document states that poor governance rather than scarcity is important and it established the basis of what is now known as the human right to water. Since then, the term water security has witnessed widespread use in academic, policy and eventually social spheres.

Some authors have restrictively linked water security to the availability of water in adequate quantities and qualities (Rijsberman, 2006; Swaminathan, 2001); others have included the importance of availability at affordable costs (Cheng, Yang, Wei, & Zhao, 2004); Grey and Sadoff (2007) added the human security dimension to the term by stating that water security should include health considerations related to livelihoods, avoiding water-related risks to people. For further reference, Gerlak and Muktharov in this issue include a comprehensive review of the history of the water security concept. Here, I wish to simply remark that this concept has remained in a state of evolution and its use has increased significantly.

A comprehensive and updated review of the concept of water security in academic debates can be found in Cook and Bakker (2013). Nevertheless, I can summarize some basic points of their findings. The authors show the increasing number of published articles on the topic in the past decade, across multiple disciplines in both policy and academic debates. The overview of the framings of water security that Cook and Bakker provide illustrates the findings of four different academic bodies of literature, each one framing water security in its own way. The first one is related to quantity and availability of water. The second body of literature concerns hazards and vulnerability of water resources (i.e. contamination, but also terrorism). The third body discusses water as a human need, and the fourth focuses on sustainable management.[3]

But, why has water security become such a fashionable concept? I can identify three initial explanations. On one hand, nation-states have been developing national security agendas with a narrow interpretation of the 'securitization' of water resources. This approach includes alarmist securitized formulations of water security as a 'threat' as states risk to be exposed to uncertain water access. The corresponding reaction is to 'safeguard the source in volumetric terms from others'. Lankford, Bakker, Zeitoun, and Conway (2013) present this problematic approach and rightfully emphasize the transnational sustainable dimension of water security: 'water security cannot be achieved at the expense of the water security of others; sustainable outcomes require reconciliation of basic needs [*particularly in border areas*] and access to water' (Lankford et al., 2013, p. 3). Consequently, water security cannot be viewed in zero-sum frameworks of political competition.

Second, as stated in Koff and Maganda (2014), water security became popular right after the widespread acceptance of the term 'human security'. With the emergence of new security paradigms, such as 'human security' and 'water security', governance debates have become more complex and more opaque. Because 'new' threats to security are transnational, regions and global governance structures, such as the United Nations (UN) have taken leading roles in defining policy responses. These responses view water as a right which all citizens must enjoy. Water security emphasizes the protection of water resources for the benefit of people, communities, and states along the same lines as human security which is defined in these terms.

Of course, persistent and dramatic challenges exist to guarantee water security. The statistics on water scarcity around the globe (also called 'water stress and/or crisis'), show that scarcity still affects about 1.6 billion people, almost one quarter of the world's population. These individuals face significant economic water shortages; or even the relevant water and sanitation resources described in the popular Millennium Development Goals (MDG) which aim to reduce the approximately 780 million individuals (one in five persons in developing countries) who do not have access to drinking water and the approximately 2.5 billion people—about half the population in developing countries—with no access to an adapted sanitation device (2012 data from World Health Organization). Because of this terrible combination, about 2 million people die every year, most of them children less than 5 years of age, from diseases associated with inadequate water supply (WHO, 2012). Indeed, water stress is an indicator linked to the availability and accessibility of water. The Falkenmark Water Stress Indicator[4] (resources to population index) is one of the most commonly used indicators when describing water availability in a country.

Nevertheless, regional differences are truly important. Even though there may be some cases where even in wealthy, water-rich countries, substantial portions of the population do not enjoy water security, the most affected populations related to the above-cited statistics are located in developing countries, where extreme water conditions—poverty conditions—remain, and water security will continue being an important concern.

Returning to my initial question, what relevance is there to talk about water security issues in country cases where water availability and accessibility do not seem to be a problem? Water security opens two interesting axes of analysis related to development:

- It is related to the capacity of a state to ensure that its inhabitants will continue to have access to potable water (which mostly happens in rich countries). Here, the concept is linked to environmental security in terms of sustainability and future water supplies.
- Water security relates to the capacity of a state to identify available, stable, and continuous access to potable water (especially in water-poor countries). Often, wealthier countries support these efforts through development aid so they become interested parties. This aspect of water security indicates stronger links to water stress and the human need for water.

For these reasons, we understand 'water security' as a normative condition anywhere in the world (poor and wealthy countries) where different populations at any level (households communities, neighborhoods, states, etc.), should enjoy access to sufficient and safe water—and its related sanitation services—to meet both their short-term and long-term needs at affordable costs for a healthy and productive life while ensuring the protection and enhancement of natural environment in local and foreign territories.

This (almost utopic) inclusive concept has been defined through normative debates at the global level,[5] meaning, water security is only valuable as a paradigm through implementation.

For this reason, it is a concept that is closely related to governance issues and therefore imperatively requires implementation capacities.

Moving toward the definition of a water-security-beyond-scarcity approach (particularly in wealthy countries), it is important to discuss the general characteristic of international norms. First, scholars of international relations (such as Goertz and Diehl 1992; Risse, Ropp, and Sikkink 1999) have noted that international norms reflect ideational commitments that go beyond the rational interests of states and other political actors. In this sense, water security cannot be appreciated simply because specific states or regions need water for their pursuits. Water security must be viewed as an idea that is characterized by inherent moral or ethical value.

Second, international norms must be viewed as universal. If states agree to uphold a norm, they cannot do so at the expense of others. Norms cannot be reserved for 'domestic consumption' (see Koff, 2009), but they must be applied in both domestic and foreign policy. Otherwise, norms lose their value as core principles that guide policy-making behavior (Manners, 2002).

In relation to water security and its relationship to wealthy states, we can identify the following characteristics that need to be respected in order for this concept to be applied as an international norm:

- Water security should be a global norm that should always privilege the needs of human beings over those of nation-states.
- Multilevel water security policies—including global, regional, national, and local governance—are needed in order to implement normative directives. Some attempts have been made, most notably at the global level in relation to the MDG 7c goal to halve by 2015 the proportion of people without access to water and the declaration of the UN Human Right to Water and Sanitation in 2010. However, in the following section, we will observe that these efforts are actually disconnected from each other. Also, they lack implementation strategies at regional, national, and local levels that would guarantee the normative infusion of 'water security' into water management policies.
- Water security policy should coordinate inter sectorial efforts, aiming to maintain sustainable and long-term water security.
- More specific to wealthy regions: water security policies should place careful attention to the risk that water security in rich countries (or regions and/or border areas) does not come at the cost of water insecurity for others somewhere else (Zeitoun, 2013). 'Water grabbing' occurs at alarming rates in all continents, except Antarctica (Rulli, Saviori, & D'Odorico, 2013). The relationship between water security and development will be analyzed in the next section. This section will also show that water security in domestic policies is a key to implementation in foreign affairs.

The following section addresses these issues through an analysis of policy-making in a wealthy region (the EU) in relation to water security debates. This regional case has double relevance: first, it implements the analysis presented above related to normative water security approaches in rich regions, and second, it illustrates how water security is operationalized in a wealthy EU member country such as Luxembourg.

3. (Lacking) Water Security Strategies in the EU

As shown above, the largest and most influential policy norms on water security are produced at the global level. For this reason, scholars of international relations recognize the universality of international norms. Unfortunately, existing regional disparities in terms of water needs, water supplies, and government structures make universal solutions nearly impossible. Thus, regional water security analysis is needed, but it also has its challenges. Wescoat and Halvorson (2012) state that 'Water governance at the regional scale reflects complex interactions and insecurities involving diverse riparians [*stakeholders*], hydro-strategic agendas, ideological dynamics, bio-physical complexities, and often competing visions of basin boundaries' (Wescoat & Halvorson, 2012, p. 87).

According to Cook and Bakker (2013), specific definitions of water security have emerged in regions where particular water security concerns are acute. For example, research on water security in Australia, China, the Middle East and North Africa (MENA) illustrates their regional specificities where disciplinary framings address particular concerns related to arid zones, availability and pollution, political and socioeconomic factors, a country's instability, and so on. For example, we can cite the efforts of Al-Otaibi and Abdel-Jawad (2007) linking a water security approach to the stability in the MENA region. Nevertheless, the literature on 'regional water security' (in terms of macro regional integration efforts such as EU, North American Free Trade Agreement (NAFTA), Andean Community of Nations (CAN), Union of South American Nations (UNASUR)) remains poorly developed and dominated by empirical studies in the above-cited regions and sub-regions where water security is linked to water scarcity.

It is very important to note that there is a dearth of literature on water security for wealthy and stable regions. This point is very important because the literature on these regions utilizes a different language. For example, the literature and policy documents related to the EU do not address water security, but they focus entirely on water governance. This is an important description because the EU and its member states are committed to water policies aimed at distribution and management without linking water concerns to security questions, like they do with other environmental issues such as climate change or food security.

One way to identify a polity's commitment to a norm is whether or not they implement the norm domestically or internationally. The EU has positioned itself in global affairs as a norm-driven actor. The EU's official foreign policy positions include the promotion of human rights and environmental security, notably in the fields of climate change and food security. Moreover, the EU promotes a policy coherence for development agenda, which contends that non-development policies must not undermine development cooperation strategies.

Having said this, closer analysis of EU human and environmental security strategies shows that the EU is more active in these arenas in foreign affairs than it is domestically. European security debates usually include human security as an essential part of the EU's development actions, but the subject is not prominent in continental politics (Koff & Nanga, 2013), and water is not the exception. The EU has codified its human security and development objectives through the Cotounou Agreements which guide development aid from the EU and its member states.[6] However, as Harlan Koff's contribution to this special issue shows, the EU has been slow to enact human security and environmental security policies domestically. Koff's article shows that EU security policies remain focused on 'hard security' issues, such as border controls, anti-trafficking measures, counter-terrorism, and efforts to combat organized crime.

This view of internal security has impeded the establishment of water security as a norm to be accepted and implemented by the EU. The research conducted through a multi-year research project on Human and Environmental Security in Border Regions: Cross-regional Perspectives (HUMENITY), funded by the University of Luxembourg, indicates that the EU has approached human and environmental security as 'products for export'. Both norms are embedded in European foreign policies but guide neither security governance nor environmental governance within the Union. This obviously affects EU policy-making in the field of water politics because water is absent from overarching security debates and in terms of environmental governance, the EU's governance strategies have been dominated by conservation approaches that do not link water resources to security or rights-based norms.

This is illustrated in Table 1, which briefly presents human security, environmental security and water security in selected cross-border regions in the EU. The table shows that these approaches have not been significantly integrated in EU cross-border policies. The two internal border regions included in this table (the Eurometropolis and Luxembourg's Greater Region) have very limited security perspectives. The external border, included in this study between Spain and Morocco is more affected by human security, environmental security, and water security because it is an area of significant human trafficking, that is also characterized by environmental degradation and water scarcity. Water security is an international issue because Ceuta and Melilla, two Spanish outposts in Africa, are dependent on Morocco for water. This table shows that the EU's commitment to these security norms is based less on ethical positions and more on strategic interests.

In fact, the EU's policies in the field of water politics clearly illustrate the difference between water governance and water security. At the regional level, the EU is widely recognized as a pioneer in environmental regime development, with important praise given for efforts to establish all institutional, normative, and juridical arrangements to establish a regional water regime. These European developments are quite recent. Since the early 1970s, the EU was conducting an extensive range of legislative measures in order to promote the sustainable management of water with policies aimed at guaranteeing both water quantity and quality in the region. A governance structure was erected that addressed both international law and transboundary cooperation. Indeed, Gerlak and Mukhtarov (2013) contend that European River Basin Organizations (RBOs) serve as models of regional water management around the world.

The current EU water regime is the result of three waves of legislation (directives) implemented in the 1970s, 1990s, and the year 2000 with the adoption of the Water Framework Directive (WFD),[7] '… an overarching piece of legislation that aims to harmonize existing European water policy and to improve water quality in all of Europe's aquatic environments' (Kaika & Page, 2003, p. 314). The WFD provides mandatory goals and a comprehensive framework for integrated and adaptive water management. Under this directive, EU Member States are not only enjoined to establish environmental objectives and ecological targets for surface

Table 1. Human security, environmental security and water security in selected cross-border cases

	Characteristics of human security	View of environmental security	View of water security
The Eurometropolis	Drug trafficking	Conservation	Water quality, sporadic drought
Luxembourg's Greater Region	No existing debate	Cattenom nuclear plant	Limited debate on price
Spain-Morocco	Human smuggling and clandestine migration	Desertification, land conservation	Water scarcity in Southern Spain, Ceuta and Melilla dependent on Morocco for water

Source: Koff and Maganda (2014).

waters in a River Basin Plan, but they are also economically responsible for the full environmental and resource costs of water services; plus the establishment of national water pricing (European Commission, 2007).

Besides its strong focus on water quality, one of the potential benefits of the WFD is a comprehensive scope toward interregional coordination of international basin district governance, economic management and water policies, and the management of groundwater. Because the European territory is considered the 'land of shared waters' (about 60% of its surface is part of a river basin shared by at least two Member States), hence the relevance of the WFD. Thus, the WFD focuses strongly on the coordination of International River Basin Districts (IRBD), and since its creation, this directive has focused on specifying how to designate IRBD territories and impacts.

Also, the WFD aims to operationalize core principles of Integrated Water Resource Management (IWRM). Specifically, the WFD aims to establish 'community-based' management structures characterized by transparency, public information, citizen consultation, and cross-border dialogues that extend beyond discussions amongst water officials, including members of civil society. The WFD not only aims to protect Europe's water resources, but it seeks to operationalize a continental water management system that is based on IWRM principles. Nevertheless, water security is not integrated in the WFD-IWRM debates.

While the WFD is a central part of European water policy, its implementation has not been easy across Europe. Ghiotti (2011) has developed an interesting study on the difficulties that many EU countries face in transposing the WFD. There are still many debates amongst the EU member states' water agencies around the production of environmental information related to this directive. Particular discussions focus on highly technical indicators and standards for water quality about which interpretation is not yet uniformly harmonized amongst water agencies and stakeholders (Timmerman & Langaas, 2004).

Amongst the environmental protection efforts, the directive established several innovative principles for water management related to groundwater, public participation in planning and the integration of economic approaches like the recovery of water service costs. In those cases where the basin extends beyond the territories of the EU, the directive encourages Member States to establish cooperation with non-Member States and manage water resources on a basin level. However, the preparation of the first River Basin Management Plans (RBMP) for the 110 national and IRBDs across Europe is still an ongoing project. The then-27 EU member states were required to establish their first RBMP by the end of 2009, including specific measures to ensure that all EU waters reached 'good' status by 2015.

With all these broad objectives beyond water quality (even though 'good status' is a key concept in the WFD), this directive is considered the tool for water reform in Europe. However, some EU Member States have been demonstrating problems and inconsistencies in different spheres of the implementation of the WFD. Some countries were just steps behind the time framework since the first stages (such as Belgium (Walloon region), Denmark, Greece, Ireland, Luxembourg, Portugal, Spain); some of them had structural concerns with the transition from old 'water autonomy/sovereignty' to new regional regulatory approach, particularly at the legal and operational transposition levels. Some countries (Belgium, Germany, Greece, Italy, Luxembourg, Portugal, and Spain) have also been called to the European Court of Justice (ECJ) and then had to redesign their national administrations and responsibilities to transpose the WFD, and/or complete their RBMP (European Commission, 2012).[8] In some of them, there are ongoing debates related to national water pricing which was also requested by the WFD. This shows a gap in understanding the effective possibilities associated with the new regulatory regime based on the EU WFD.

The aforementioned analysis illustrates that the EU has not regionally implemented water security on the continent. This has also affected the EU's commitment to water security internationally. Despite a clear compatibility between water security and the Cotonou Agreement's guiding norms, the EU has not supported water security as a normative approach in its development aid policies.

In a recent paper presentation in the Nordic Development Conference in Helsinki, I have presented research on EU development cooperation strategies that examined how compatible the EU's policies are with 'the human right to water'. The presentation showed that the EU is implementing development strategies that support and adhere to the MDG. These goals, however, are indicator-based and they are not transformative (see European Reports on Development 2011/ 2012 and 2013). This approach measures water security simply in terms of quantitative measures related to access. It does not at all address ethical or normative issues related to water security.

The EU, thus far, has not supported rights-based approaches to water governance in its foreign affairs. The policies presented by the EU-Development and Cooperation (EUROPEAID) website demonstrate this. The EU Water Initiative (EUWI) highlights its role as international political initiative that 'mobilizes all available EU resources and coordinates them to achieve the water-related MDGs in partner countries'. Like the aforementioned European Water Directive which approaches water governance as a technical issue, this initiative marginalizes the role of 'rights' in water management discussions. This approach guides European development policies which often focus on technical obstacles to water distribution in poor areas. Because water is a critical resource for socioeconomic development and because water policy is an important tool in the fight against or reinforcement of development inequalities in the world, this strategy is especially relevant for human security concerns in Europe and beyond. In this respect, the EU demonstrates that its water management strategies are characterized by domestic–foreign policy coherence. However, this cannot be considered 'normatively coherent' (see Koff, 2013) as water governance approaches in both domestic and foreign policy do not reinforce the EU's declared commitments to human rights.

This incoherence is plainly clear since 2010 when the UN recognized the human right to water and sanitation. This initiative was forwarded in the UN by South American countries, most notably Bolivia. It became an issue that developing states in different world regions could rally around, which is why the initiative passed. When this right was voted upon by the UN general assembly in July 2010, the EU's member states all abstained, thus contributing to the resolution's failure to pass in the assembly (the resolution was adopted by consensus by the Human Rights Council in September 2010). While the EU remains committed to reinforcing access to water as good management practice and it is committed to improving access to water as part of its anti-poverty programs, it does not recognize the right to water nor water security positions because of their normative implications.

Much of the literature on water security (Zeitoun, Warner) contends that the needs of vulnerable populations are not met because of a lack of water security policies and strategies. Following this logic, one would expect water security policies to emerge in the European case, as it is one of the most institutionalized regional water regimes in the world. Why then, do we witness the lack of attention to water security in the European regional integration effort? Some scholars (such as Hall, 2006) have contended that this rights-based approach would countervail the interests of important European multi-national corporations, such as Suez or Vivendi which are promoting privatization strategies in water management throughout the world. However, the lack of support for European water security policies also comes from positions taken by EU nation-states, especially those where water scarcity is not an issue. This is the focus of the next section, which examines the relationship between water security and water management practices in Luxembourg.

4. Luxembourg's National Water Governance and Development Strategies: Different Actors, Same Approaches

The previous section has shown that the EU has not adopted water security norms in either its domestic or foreign water governance strategies. The EU's position, however, reflects its position as a 'water-rich' region of the world. Few European states, Spain, Portugal, and, for instance, all Mediterranean countries, contend with structural or incidental drought stress. This significantly affects the EU's position in this field. It shows that EU policy-making in this arena reflects the interests of the organization's member states, almost all of which are 'water-rich', and that water security discourse in these countries may not translate into water security practice.

One of the countries which has been slow to adopt international water management norms is the Grand Duchy of Luxembourg, which is one of the smallest countries in Europe characterized by high economic and demographic growth rates. The country has experienced dramatic changes related to population, markets, and infrastructure over the last 20 years, since its transformation from basic agriculture and heavy industries (steelworks) to a service-based economy focused on EU institutions and banking.

Like many states, immigration has played a key role in Luxembourg's economic success. From 1990 to 2009, Luxembourg's population has increased at an average annual growth rate of 1.4% dominated by immigration (40% of the country's population is foreign-born), and this trend will likely persist in the future. However, unlike most advanced industrial states, Luxembourg relies heavily on commuters to contribute to the country's economic well-being. Every day, Luxembourg receives more than 120,000 cross-border workers which is significant for a country with a population of 511,840[9] in January 2012 (see Figure 1).

For this reason, most observers (see Evrard & Shulz, 2015; Sohn, Reiter, & Walther, 2009), both academics and practitioners, have now focused their attention on the polycentric Greater Region in which Luxembourg is situated (see Figure 2). The Greater Region includes Luxembourg and the regions of Lorraine (France), Wallonie (Belgium), the German-speaking Community of Belgium, Rhineland-Palatinate (Germany), and Saarland (Germany). The area covers 65,401 km[2], its population includes 11.2 million inhabitants and it has been funded and politically recognized by the EU through its INTERREG programs since 2000.[10]

Because Luxembourg is a small state which is heavily integrated in an officially recognized cross-border region,[11] the whole country is impacted by cross-border interactions on a daily basis, including the flows of four Transboundary Rivers (Moselle, Sûre, Our, Alzette).[12] Access to water is fundamental to support the increasing population and economic growth cited above. Water supply does not seem to be a problem for Luxembourg. According to the Luxembourgish Water Administration Agency (Administration de la Gestion de l'Eau—Luxembourg), the national drinking water demand/consumption is around 120,000 m^3 per day. About two-thirds of this amount is provided by groundwater and a third comes from treatment of surface water.[13] The surface waters, particularly the international rivers mentioned above, still play important roles in the irrigation of agriculture, commercial navigation, tourism, official border delimitations. They even play a role in the unification of Europe, such as their symbolic place in the formulation and signing of the Schengen Accords.[14] Thus, Luxembourg is a country surrounded by water and this resource significantly impacts its political, cultural, and economic life. Because Luxembourg's economic well-being is linked to cross-border territories and their water resources, it can be assumed that water security should be a priority for this country.

Despite this fact, an analysis of Luxembourg's national water governance strategies demonstrates a clear lack of attention to the concept of water security. This is evident in Luxembourg's national water policies. Specifically, the country has been very slow to implement European directives in the field of water governance, particularly the European WFD which is an initiative aimed at pollution control through international cooperation in transboundary river basins.

Before the WFD was enacted, Luxembourg had no national water agency. Water policy-making was conducted in a piecemeal fashion with representatives of related ministries conducting ad hoc discussions on immediate issues related to water quality and water pricing. The country sent representatives from different ministries to participate in cross-boundary basin councils related to the Rhine and Meuse rivers on a rotating basis.

The WFD, which was passed in 2000, was supposed to force countries like Luxembourg to broaden their water policies in accordance with the principles of integrated water resources management (IWRM). However, this has been slow to occur. Luxembourg did not establish a national water agency until 2005 and it did not transpose the WFD into a national water law until 2009, six years after this was supposed to have been accomplished. Many observers claim that this occurred only because Luxembourg was forced to do so (see Maganda, 2013). In fact, in 2006, the ECJ sided with the European Commission in a case in which Luxembourg contended that the WFD required 'IWRM' and not an 'integrated water law'. The ECJ's actions essentially forced Luxembourg to institutionalize its water policies or face significant economic sanctions.

While Luxembourg did pass this legislation, implementation has been slow. For example, according to an expert on water governance in the Greater Region (interview conducted by author):

> ... You also have to know that the WFD is extremely demanding in terms of its conceptual design, as well as the reasons why there are delays. There is what I call the 'First Round' which is the first cycle from 2000 until 2015, and then there will be a second and a third cycle, and it will go on and on. The first cycle will probably not be very successful in terms of implementation but what will be gained is that there has been a huge effort to try to make all different instruments work, and the agenda of the WFD totally overestimated the technical and scientific means that were there to implement that. Nobody was ready for that, starting with the biology where the different [biological] indices were not really ready for the actual assessment of the ecological status. Nobody really knew what that would mean, the ecological status.

Luxembourg's most recent case in front of the ECJ results from the country's inability to implement the WFD. The aforementioned 2006 verdict against Luxembourg called for the country to refurbish 12 water treatment plants. Luxembourg did not begin this refurbishment until 2011 when the European Commission threatened legal action. In November 2013, the ECJ ruled in favor of the Commission and fined Luxembourg two million euros for not having complied with its previous verdict. Moreover, the court fined Luxembourg 2800 euros per day in which the country does not comply.

These actions by the ECJ demonstrate that Luxembourg approaches water governance as a management issue. In fact, because water supply is abundant in the region, little public discussion has evolved over water issues and there is little social participation in water politics. EU Eurobarometer reports have also confirmed the weak public participation in WFD

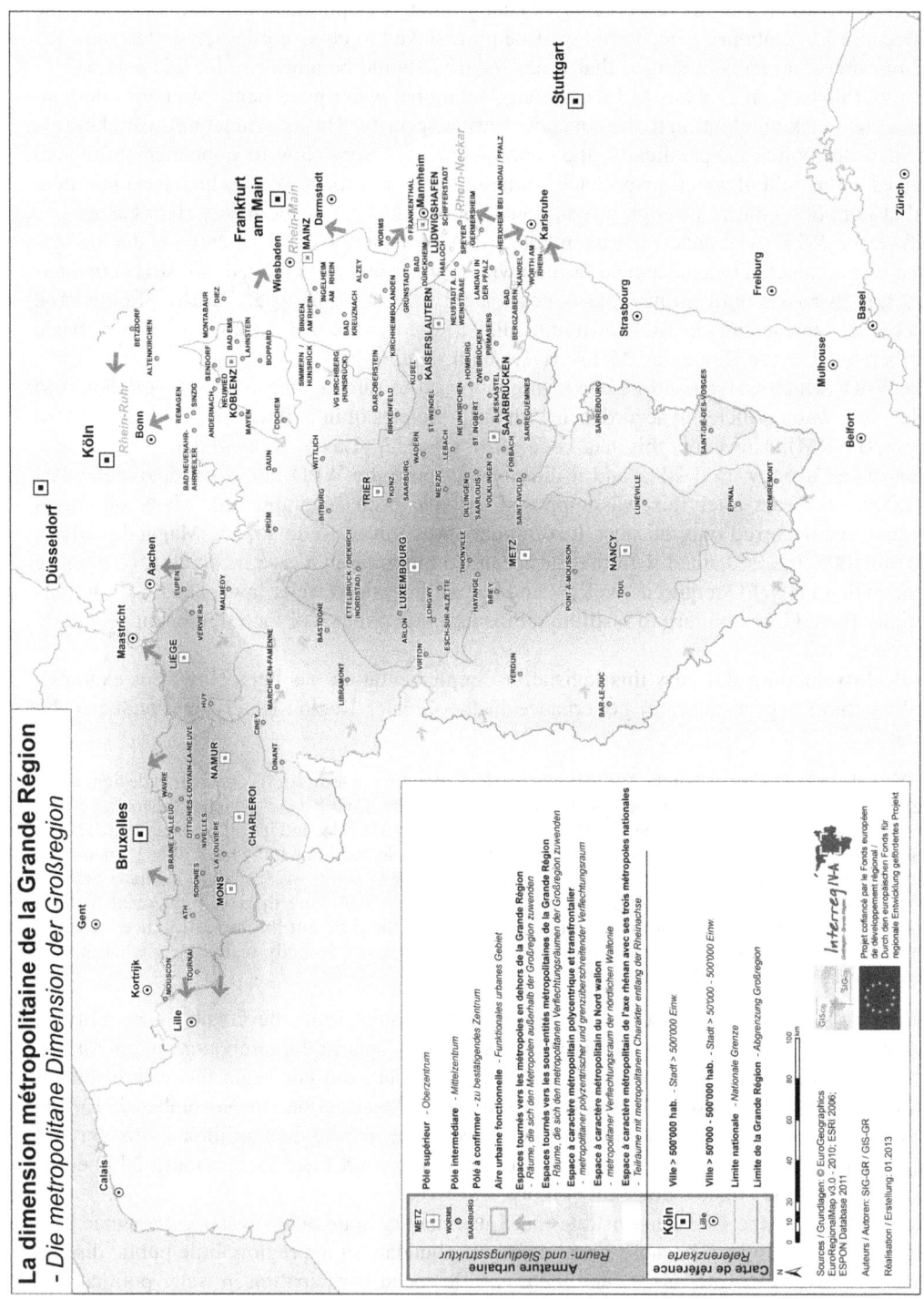

Figure 1. The metropolitan dimension of the Greater Region.

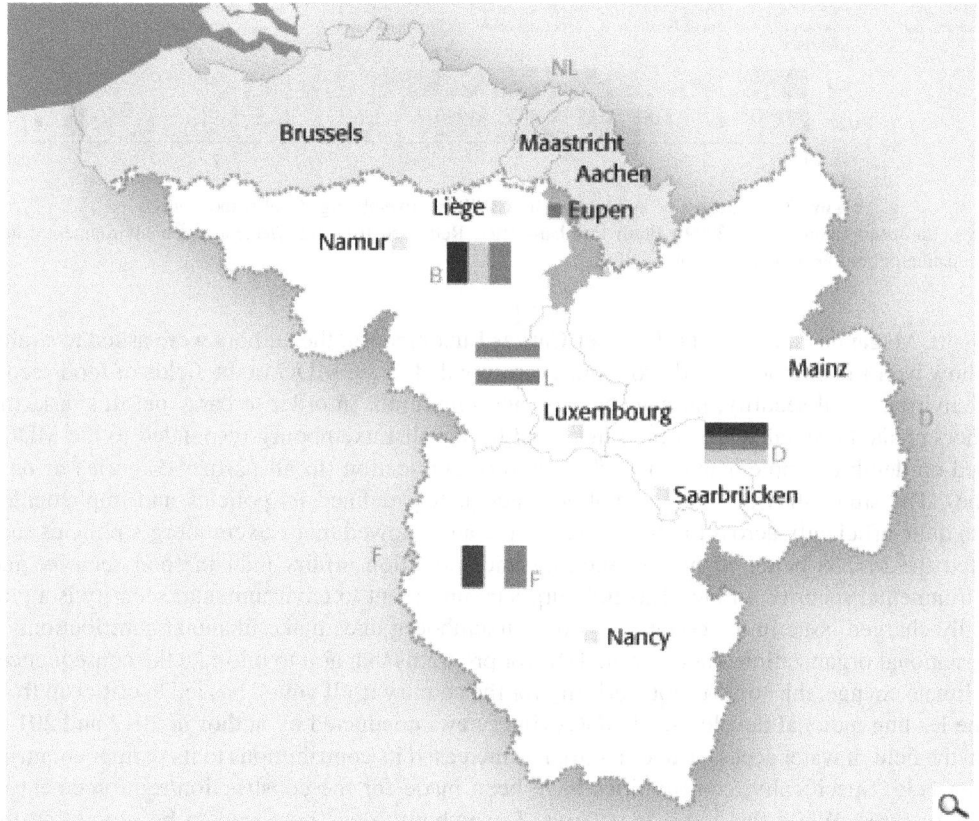

Figure 2. The Greater Region.
Source: Map designed by Jose Luis Alvarez Palacios (INECOL), based on map provided at http://www.go-east.be/de/business/standort/saarlorlux.html (consulted 20 January 2016).

implementation in Luxembourg, thus supporting the tendencies indicated by national reports and interviews. In response to the following question:

> The EU requires Member States to prepare a plan for the management of water resources to achieve good water quality by 2015, these are the River Basin Management Plans, and to consult the public and interested parties in this process. Are you aware of a consultation by the authorities on the river basin management plans where you live? And are you planning to express your views in this consultation?

Roughly one-fifth of Luxembourgish respondents are aware of the consultation process (19% vs. 12% in the EU), while approximately 8 in 10 are not (78% vs. 76% in the EU). Two percent of respondents had already participated in the process and about half have the intention to do so (51% vs. 50% in the EU) (see Figure 3).

Given this domestic scenario, it is not surprising that Luxembourg has not adopted water security norms in its foreign policies either. Luxembourg is a recognized actor in global development politics as it allocates 1% of its GDP to development aid. Moreover, it works with 10 selected partner countries and it is recognized for the durability of its aid commitments.

Having said this, Luxembourg's development aid programs are designed within the framework of technical support and they are not necessarily transformative. In a study conducted

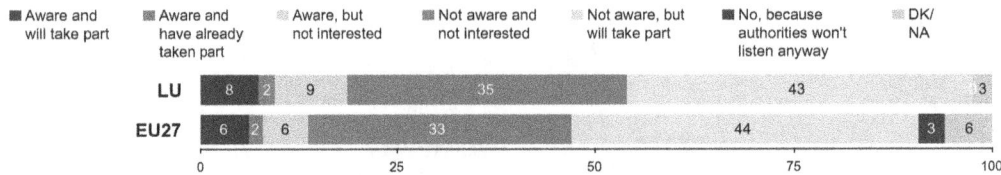

Figure 3. Awareness of and interest in RBMP in Luxembourg (% of respondents).
Source: European Commission (2009). Flash Eurobarometer. Retrieved from http://ec.europa.eu/environment/water/participation/pdf/eurobarometer_summary.pdf

by Koff, Maganda, and Nanga (2013) for Caritas Luxembourg, the authors were asked to evaluate how well Luxembourg's aid programs corresponded to the MDG in the fields of food security, environmental security, human security, and migration. In order to carry out this task, the authors created ordinal scales which indicated how well Luxembourg responded to the MDGs based on depth of implementation and breadth of application (to all partner countries or only some). The study showed that Luxembourg adequately defined its policies and implemented them quite efficiently across the board. The study also showed that Luxembourg's policies supported the MDGs better in human security and migration affairs than in food security and environmental security. In fact, Luxembourg's commitment to environmental security is a politically charged issue in the country. While, Luxembourg does make financial contributions to international organizations (notably the UN) for programs that aim to mitigate the consequences of climate change, this topic is not a priority for the country itself which has led to criticism from some leading national development NGOs (interviews conducted by author in 2012 and 2013).

In the field of water access, Luxembourg has increased its contributions to its partner countries in the field. Specifically, contributions have been made for the construction/reinforcement of infrastructures. While this trend is positive, Luxembourg does not seem to be promoting the water security norm presented in Section 3 above. The study for Caritas did not only study Luxembourg's investments and policies, but it also examined the nature of collaborations with the partner countries through policy documents and interviews. The study found that the weakest part of Luxembourg's development cooperation in the field of water management was the lack of aid recipient participation in the formulation of development objectives. This participation should be a fundamental aspect of a rights-based water security approach. It is not surprising, however, that it is not prioritized by Luxembourg in its development cooperation strategies, given that social participation is not promoted domestically. Like the EU, Luxembourg's domestic and foreign policy water strategies are characterized by coherent approaches. However, like all other EU member states, Luxembourg also abstained in the 2010 vote on the human right to water in the UN general assembly. Thus, despite a declared commitment to international development and human rights, we can conclude that Luxembourg has not infused its water governance strategies with these norms, thus contradicting a commitment to the notion of water security.

5. Conclusions

Water security has emerged as an impressive norm in global affairs. The human right to water and sanitation is a transformative paradigm which has the potential to radically shift how we manage water resources at the local, national, and supranational levels. The problem with this norm, however, is the implementation. Like the broader concepts of human security and environmental security, the breadth of these new norms makes it very difficult to establish mechanisms and rules for implementation.

One of the most interesting aspects of water security as an evolving international norm is its potential for transformative policies. Thus far, development cooperation has been guided by the MDG's framework. The MDGs of course, are indicator-based rather than transformative in nature. While they are important because they established benchmarks for human security and development objectives, the MDGs have not promoted new governance relationships either in global affairs or in water management. As stated in the introduction above, access to safe water and sanitation has been measured quantitatively in terms of infrastructure and water distribution. However, there is no mention of addressing power relationships or the displacement of environmental problems across space and time by wealthy countries.

Water security, as it has been defined above, would directly address these problems. By framing water governance in terms of rights and responsibilities, instead of the efficiency/effectiveness of water delivery, the right to water would become a pillar of human security and human development strategies throughout the world. Instead of debating whether water should be privatized, political debates could focus more on what wealthier polities should do in order to support access to water in developing states/regions, as also suggested in the analysis of water security presented in the introduction.

In fact, water security is recognized as an international norm, but what value does this have in global affairs? This article has argued that if a norm has truly been adopted by a state/region, then it should guide policies in both domestic and international affairs. Thus far, the notion of water security has been adapted and operationalized in many developing states/regions. These polities (such as Bolivia/Andean Community of Nations) coherently implement water security strategies in their domestic systems and they promote the human right to water internationally, even though, as Perrault (2006) has shown, the successful right-to-water revolt in Bolivia did not impede imbalances for who gets the water now.

If a norm is to be universally accepted, it needs to be adopted by wealthy and powerful countries/regions that are 'water-rich'. If these polities do not support this norm, then operationalization becomes a serious challenge. Having said this, one would expect water-rich areas to support the human right to water because most countries in these regions are liberal democratic states that promote human rights in global affairs.

This article has shown that this expected outcome is not accurate. The EU and its member states have not supported water security-based governance strategies. For unclear reasons, the EU has not indicated any intention to support the human right to water approach proposed by Bolivia, Ecuador, and Venezuela. Questioning why would be worth pursuing further in another article.

The analysis of the WFD presented above indicates that the EU only adopts water security frameworks in localized zones of water scarcity and in the improvement of regional water quality. Moreover, EU development cooperation has focused on technical contributions to water access (i.e. drilling wells, installing pipes, etc.) without addressing the underlying causes of regional inequalities related to access to water. The member states of the EU have generally adopted a similar position. This article examines current water management strategies in the Grand Duchy of Luxembourg. Like the EU, Luxembourgish authorities have ignored water security issues both domestically and internationally. Moreover, local authorities seem to pursue policy objectives that maintain their own influence within domestic political structures rather than implementing international norms.

This perspective seems to verify a recurring claim in global affairs: wealthy regions do not support water security or the human right to water through domestic or international policy strategies. This article claims that a principle cannot be considered an international norm until it has been adopted by wealthy and water-rich states. If only countries characterized by water scarcity support water security strategies, then the concept loses some of its moral strength as it reflects rational incentives based on material needs.

Moreover, water security debates reflect power imbalances in global affairs. Scholars of regional and cross-border water governance have traditionally argued that water management is strongly influenced by power imbalances as wealthy regions can affect access to water resources in their favor to the detriment of the populations living in poorer areas. This is the reason why indicator-based water management strategies cannot guarantee water security at the international or regional levels. For water security to be implemented as an international norm, political and academic attention should focus more on wealthy, water-rich states, which have often been ignored in these debates. Obviously, regional power differences will remain but, should water-rich states such as Luxembourg begin to adapt and implement human security norms, then transformative change can occur over time through accumulation of national measures. Without the support of wealthy regions/countries, water security cannot emerge as a transformative paradigm.

Disclosure Statement

No potential conflict of interest was reported by the author.

Notes

1 According to www.eea.europa.eu/soer, over the period 1995–2009, the economy grew strongly, with GDP at constant price increase by 75.6% for an average of 4% per year. When looking at GDP per capita and the purchase power, Luxembourg becomes the richest country in the EU as well as in Organization for Economic Co-operation and Development (OECD)—the highest GDP in the OECD countries during 2010 and 2011. Its gross value added is largely generated in the financial and corporate service sector, but also reflects the importance of cross-border workers. Retrieved from http://www.eea.europa.eu/soer/countries/lu/soertopic_view?topic=freshwater and http://www.oecd.org/dataoecd/33/43/49655399.pdf (consulted March 28, 2012).

2 Understanding water governance as the political, social, economic, and administrative systems that are in place, and which directly or indirectly affect the use, development, and management of water resources, and the delivery of water service at different levels of society (taken from the Water Governance Facility program of the United Nations Development Program). Retrieved October 10, 2014, from http://www.watergovernance.org/whatiswatergovernance

3 Further references of related literature can be found in Cook and Bakker (2013). Some other reflections also address the relationship of water security with the Integrated Water Resources Management, and Food Security.

4 The Falkenmark Water Stress Indicator was developed by the Swedish water expert Falkenmark in 1989. Retrieved from http://environ.chemeng.ntua.gr/WSM/Newsletters/Issue4/Indicators_Appendix.htm

5 Amongst the most relevant water security fora that address this issue, one finds the GWP, World Water Forum, UN-Water, World Economic Forum (addressing water security since 2008), and United Nations Educational, Scientific and Cultural Organization- Institute for Water Education (since 2009 water security is one of the research axes of their Institute for Water Education). The UN finally established access to water and sanitation as a human right in 2010.

6 The Cotonou agreement which guides European development cooperation strategies outlines four main principles: equality of partners and ownership of development strategies, participation, dialogue and mutual obligations, and differentiation and regionalization.

7 After an interesting history of conflict-driven policy-making including parliamentary debates, strong participation of environmental NGOs and political environmental lobbies, the WFD was ratified and it entered into force in

December 2000 with the aim of guaranteeing safe water and controlling pollution of freshwater sources. EU member states had three years from the adoption date to transpose the directive into national legislations.

8   Commission report to the European Parliament and the Council on the implementation of the WFD—RBMP (COM(2012)670 of 14.11.2012).

9   Le Portail des Statistiques Grand-Duché de Luxembourg. Retrieved from http://www.statistiques.public.lu/fr/actualites/population/population/2011/05/20110503/index.html (consulted April 3, 2012).

10 http://www.granderegion.net/fr/grande-region/index.html

11 About 82 km long and 57 km wide, with borders easy to reach within an hour's drive (sometimes less): to the North-East, the German Bundesländer of Rhineland-Palatinate; to the East, the German Saarland; to the south, the French region of Lorraine and to the West and North, the Belgian Walloon Region.

12 There are four large rivers crossing Luxembourg and they all have cross-border origins.

- The Moselle originates in France at the western slope of the Ballon d'Alsace in the Vosges mountains, and flows through the Lorraine region further downstream drawing 42 km of the Luxembourgish–German border. It flows into the Rhine at the level of Koblenz (Germany). Total length: 545 km. A small part of Belgium is also drained by the Mosel through the Our.

- The Sûre (Sauer) is a Belgium-origin river rising near Vaux-sur-Sûre in the Ardennes. It forms the border between Belgium and Luxembourg for 13 km north of Martelange, then toward Germany. A left tributary of the Moselle River, its total length is 173 km. After flowing through Ettelbruck and Diekirch, the Sûre forms the border between Luxembourg and Germany for the last 50 km of its course, passing Echternach before emptying into the Moselle in Wasserbillig. The rivers Wiltz, Alzette, White Ernz, Black Ernz, Our, and Prüm are tributaries.

- The Our has its origins in the Hautes Fagnes, south-eastern Belgium. It flows southwards oriented along the Belgian–German border and after Ouren along the German–Luxembourgish border for about 34 km, ending as left tributary to the river Sûre(Sauer) in Wallendorf. The historic city Vianden lies along the Our. Its total length is 78 km.

- The Alzette rises in Thil, part of the Meurthe-et-Moselle French département. It crosses the border between France and Luxembourg flowing through the Luxembourgish towns Esch-sur-Alzette, Luxembourg City and Mersch. It delivers into the Sûre(Sauer) near Ettelbruck. It has a total length of 73 km.

    Information compiled by the author based on web pages of International Commission for the Protection of the Mosel and Saar, Meuse Hydrographic District and Wikipedia-Luxembourg.

13 Groundwater is captured by about 300 springs and wells fed mainly by water from the sandstone aquifers of Luxembourg and the Buntsandstein area. Drinking water from surface water is produced at the dam of Esch/Sure. This production requires a complex treatment, performed by the Syndicat des Eaux du Barrage d'Esch/Sure (SEBES). The maximum capacity of treatment plant is 60,000 m^3/day. Besides the treatment of surface water, the groundwater SEBES exploits the captures by drilling at three different sites. Retrieved from http://www.eau.public.lu/eau_potable/production_distribution_responsabilites/index.html, consulted March 29, 2012.

14 The Schengen Accords, or Schengen Agreements, are amongst the pioneer treaties of the EU, related to the freedom of movement and no internal border controls. The first agreement was signed on 14 June 1985 near the town of Schengen in Luxembourg, between 5 of the 10 member states of the European Economic Community. The definition of the 'Schengen area' is representative of a territory where the free movement of persons is guaranteed. Today, Europe's borderless zone, currently covers 26 European countries, with a population of over 400 million people in an area of 4,312,099 km^2 with no internal border controls.

References

Al-Otaibi, A., & Abdel-Jawad, M. (2007). Water security for Kuwait. *Desalination, 214*, 299–305.

Cheng, J., Yang, X., Wei, C., & Zhao, W. (2004). Discussing water security. *China Water Resources, 1*, 21–23.

Cook, C., & Bakker, K. (2013). Debating the concept of water security. In B. Lankford, K. Bakker, M. Zeitoun, & D. Conway (Eds.), *Water security: Principles, perspectives and practices* (pp. 49–63). Earthscan water text. London: Routledge.

European Commission. (2007). *Communication from the commission to the European parliament and the council towards sustainable water management in the European Union—First stage in the implementation of the Water Framework Directive 2000/60/EC.*

European Commission. (2009). *Flash EB Series #261 Flash Euro barometer on water conducted by The Gallup Organisation, Hungary upon the request of Directorate General Environment.*

European Commission. (2012). *Report on the implementation of the water framework directive (2000/60/EC).* DG Environment-WFD.

European Unión. (2011/2012). *The 2011/2012 European Report on Development, Confronting Scarcity: Managing Water, Energy and Land for Inclusive and Sustainable Growth.* Overseas Development Institute (ODI), European Centre for Development Policy Management (ECDPM), German Development Institute/Deutsches Institut für Entwicklungspolitik (GDI/DIE), Brussels.

European Unión. (2013). *The 2013 European Report on Development, Global Action for an Inclusive and Sustainable Future.* Overseas Development Institute (ODI), European Centre for Development Policy Management (ECDPM), German Development Institute/Deutsches Institut für Entwicklungspolitik (GDI/DIE), Brussels.

Evrard, E., & Schulz, C. (2015). L'ambition métropolitaine: clé vers un aménagement du territoire transfrontalier en Grande Région SaarLorLux? *L'information géographique, 79*(3), 54–78.

Gerlak, A., & Mukhtarov, F. (2013). River Basin Organizations in the global water discourse: An exploration of agency and strategy. *Global Governance, 19,* 307–326.

Gerlak, A. K., & Wilder, M. (2012). Exploring the textured landscape of water insecurity and the human right to water. *Environment, 54*(2), 4–17.

Ghiotti, S. (2011). La directive cadre sur l'eau (DCE) et les pays méditerranéens de l'union européenne. Une simple question de ressources en eau? *Pôle Sud, 2011/2*(35), 21–42.

Goertz, G., & Diehl, P. (1992). Toward a theory of international norms. Some conceptual and measurement issues. *The Journal of Conflict Resolution, 36*(4), 634–664.

Grey, D., & Sadoff, C. W. (2007). Sink or swim? Water security for growth and development. *Water Policy, 9*(6), 545–571.

Hall, D. (2006). Corporate actors: A global review of multinational corporations in the water and electricity sectors. In D. Chavez (Ed.), *Beyond the market: The future of public services* (pp. 179–185). Public services yearbook (2005/6). London: TNI/Public Services International Research Unit (PSIRU).

Kaika, M., & Page, B. (2003). The EU water framework directive part 1: European policy-making and the changing topography of lobbying. *European Environment, 13*(6), 314–327.

Koff, H. (2009). Creating exclusion through integration strategies: The impact of gypsy policies in Western Europe. In H. Koff (Ed.), *Social cohesion in Europe and the Americas: Power, time, and space* (pp. 85–113). Bruxelles: P.I.E.-Peter Lang.

Koff, H., & Maganda, C. (2014). Water security in cross-border regions: What relevance for regional human security regimes? In D. Garrick & G. Anderson (Eds.), *Water security and federal rivers* (pp. 325–338). Cheltenham: Edward Elgar Publishing.

Koff, H., Maganda, C., & Nanga, E. (2013). *Rapport sur l'action du Luxembourg dans le cadre de l'achèvement des objectifs du millénaire pour le développement* (RISC Consortium Working Paper # 6) pp. 1–72. Luxembourg: Online publication. Retrieved from: http://www.risc.lu

Koff, H., & Nanga, E. (2013, November). *Coherence for development for whom?: Examining the development-migration nexus in Morocco and Mexico.* Paper presented at knowing development—Developing knowledge? 2nd Nordic conference for development research, Helsinki, pp. 14–15.

Lankford, B., Bakker, K., Zeitoun, M., & Conway, D. (2013). *Water security: Principles, perspectives and practices.* Earthscan water text. London: Routledge.

Maganda, C. (2013). The implementation of the European water framework directive in Luxembourg: Regional compliance vs. cross-border cooperation? *International Journal of Water Governance, 1*(3–4), 403–426.

Manners, I. (2002). Normative power Europe: A contradiction in terms? *Journal of Common Market Studies, 40,* 235–258.

Perrault, T. (2006). Resource governance, neoliberalism and popular protest in Bolivia. *Antipode, 38,* 150–172.

Rijsberman, F. R. (2006). Water scarcity: Fact or fiction? *Agricultural Water Management, 80*(1–3), 5–22.

Risse, T., Ropp, S., & Sikkink, K. (1999). *The power of human rights: International norms and domestic change.* Cambridge: Cambridge University Press.

Rulli, M. C., Saviori, A., & D'Odorico, P. (2013). Global land and water grabbing. *Proceedings of the National Academy of Sciences of the United States of America. 110*(3), 892–897.

Scott, C. A., Varady, R., Meza, F., Montaña, E., de Raga, G. B., Luckman, B., & Martius, C. (2012). Science-policy dialogues for water security: Addressing vulnerability and adaptation to global change in the arid Americas. *Environment: Science and Policy for Sustainable Development, 54*(3), 30–42.

Sohn, C., Reiter, B., & Walther, O. (2009). Cross-border metropolitan integration in Europe: The case of Luxembourg, Basel, and Geneva. *Environment and Planning C: Government and Policy, 27*(5), 922–939.

Swaminathan, M. (2001). Ecology and equity: Key determinants of sustainable water security. *Water Science and Technology, 43*(1), 35–44.

Taylor, R. (2009). Rethinking water scarcity: The role of storage. *Eos, Transactions American Geophysical Union, 90*(28), 237–238.

Timmerman, J. G., & Langaas, S. (Eds.). (2004). *Environmental information in European transboundary water management.* London: IWA Publishing.

Wescoat, J., & Halvorson, S. (2012). Emerging regional perspectives on water research and management: An introductory comment. *Eurasian Geography and Economics, 53*(1), 87–94.

World Health Organization (WHO). (2012). *Progress on drinking water and sanitation.* Report 2012—Update. Retrieved from http://www.who.int/water_sanitation_health/publications/2012/jmp_report/en/index.html

Zeitoun, M. (2013). The web of sustainable water security. In B. Lankford et al. (Eds.), *Water security: Principles, perspectives and practices* (pp. 11–25). Earthscan water text. London: Routledge.

Scarcity and Power in US–Mexico Transboundary Water Governance: Has the Architecture Changed since NAFTA?

STEPHEN P. MUMME

ABSTRACT   *This paper examines the politics of water allocation on the US–Mexico border since the North American Free Trade Agreement (NAFTA) began in 1994. While NAFTA reforms have modestly changed the water allocation regime, they have not altered the longstanding asymmetry of power relationships governing the allocation of water resources between the two countries. Two rivers are considered. On the Rio Grande, NAFTA and its associated reforms had the effect of accentuating recent allocation crises and helping to resolve them, while leaving existing power arrangements largely intact. On the Colorado River, efforts to save the Colorado River delta ecosystem after NAFTA benefitted from institutional reforms, but these efforts remain rather marginal to the longstanding structure of power governing allocation and management of Colorado River water resources, as the case of the All-American Canal dispute reveals. These cases reveal the treaty regime as one that is highly resistant to change, suggesting that caution is needed when using theoretical constructs like multilevel governance and collaborative watershed management in drawing generalizations on transboundary water management along the US–Mexican border.*

Introduction

To what extent, if any, has the North American Free Trade Agreement (NAFTA) altered the architecture of power in US–Mexico water relations? At a time of chronic water shortage across the transboundary river basins that serve the two countries, this is not just an academic question. It gets directly to the capacity of the governments to equitably address pressing water needs along the US–Mexico border, an arid region that continues to rapidly develop and generate a host of competing demands on the area's limited water resources. Transboundary water governance in the face of acute and chronic shortage arguably requires greater binational cooperation if conflict is to be avoided and resources are to be managed sustainably to the general satisfaction of stakeholders on both sides of the border. The NAFTA reforms, particularly its environmental side agreements and related developments, held out the prospect of greater cooperation on a range of natural resources and environmental concerns including the management of water resources. Has NAFTA made a difference in transboundary water governance and, if so, how?

This paper explores these questions. It proceeds, first, by retracing the basic architecture of US–Mexican water relations and reviewing several schools of thought on how the NAFTA reforms may have altered the water governance equation in transboundary river basins along the US–Mexico border. It follows with a brief review of important developments and recent problems in the management of river water in the two major river basins that dominate most discussions of US–Mexico water relations. These two cases and their subcases reveal important changes in water governance over the past 20 years that have generated some guarded optimism concerning the prospects for greater binational cooperation related to water management in these important basins. The paper concludes by reflecting on the impact of these changes on the overall structure of the bilateral relationship for transboundary rivers management along the border.

Structure and Reform in Transboundary Water Management: Theoretical Perspectives

In a recent article coauthored with Stephen Mumme and Oscar Ibanez (2013, p. 169), we argued that 'the structural underpinnings of power in US–Mexico water management are profound and enduring'. This is certainly the case. The allocation and management of transboundary water has historically evolved under circumstances of considerable economic and strategic asymmetry since the boundary was set in 1848, conditions that have not fundamentally changed even as the two countries have moved toward greater economic integration after WWII. Water relations are also affected by sovereign constitutional and political differences that shape the political dynamics of resolving water disputes (Mumme & Ibanez, 2013). With water governance centered at the national level in Mexico, and the state level in the USA, for example, the difficulty of achieving binationally acceptable agreements is considerable. Such differences accentuate the importance of agreements when achieved and raise the stakes associated with their undoing. Regional water scarcity has ensured that these agreements are foundational to national development on both sides of the border, accounting for the fact they rank among the earlier and more important of the national agreements struck by the two countries. In fact, the US–Mexico relationship on transboundary water is from a diplomatic point of view one of the most fixed arrangements in their common affairs, evolved over 165 years since the Treaty of Guadalupe Hidalgo and constitutionally determined by the 1944 Water Treaty and earlier agreements affecting the management of water on the Colorado and the Rio Grande Rivers. I say constitutionally determined because, in fact, the 1944 Water Treaty is one of the rarest of treaties, a 'constitutional treaty'. Such constitutional treaties, I argue, share a slate of basic political characteristics wherever they are found (see Table 1).

This deep institutionalization of transboundary water management, administered through the International Boundary and Water Commission (IBWC) and subject to the oversight of the national foreign ministries, has a different sovereign expression at the domestic level of each country that is reflected in national water administration and the politics of water affecting the governance of their transboundary rivers. At the international level, however, binational

Table 1. Criteria for constitutional treaties

- Linked to well-institutionalized national constitutions
- Linked directly to individual property rights and/or pocketbook issues—deep penetration across civil society
- Political costs of textual change are extremely high. Subject complexity that raises stakes of any alterations
- Implementation change strictly through interpretation and political agreements on amendments.
- Treaty survives multiple political administrations in both countries
- Subnational jurisdictions have high investment in treaty
- Treaty widely associated with national pride and popular conceptions of sovereignty
- Treaty core unmodified for at least 50 years

Source: Author's analysis of treaties listed in the International Freshwater Treaties Database, Oregon State University, available at: http://www.transboundarywaters.orst.edu/database/interfreshtreatdata.html.

water management has been highly predictable, grounded in treaty interpretation and long-established diplomatic practice. Its purpose and focus have historically centered on the administration of treaty rules and procedures related to water allocation, whose priority prevails over other important values embedded in the treaty regime, or without.[1]

This established institutional architecture prevailed unchallenged until the 1980s when a new agreement, the La Paz Agreement on border region environmental management, was signed by the governments. The 1983 La Paz Agreement opened an institutional window for the consideration of water quality and biodiversity issues that had been previously neglected by the IBWC, drawing new domestic and international players into transboundary rivers diplomacy. It also raised domestic awareness of border area environmental conditions that played into the politics surrounding NAFTA a nearly a decade later. The NAFTA debate in its turn produced a mix of new programs and international agencies that were additive to the extant water management regime along the border.

There is no question that the NAFTA side agreements and related programs altered the institutional environment for transboundary water management along the border. They established three new international agencies, the Commission for Environmental Cooperation, the Border Environment Cooperation Commission (BECC), and the North American Development Bank (NADB). They strengthened implementation of the La Paz Agreement along the border. They amplified natural resources policy cooperation through the *Trilateral Committee on Wildlife*. New domestic advisory bodies, focused on sustainable development and environmental protection on each side of the border, were established. The political process associated with NAFTA proved a catalyst for NGO engagement and network along and across the border, strengthening the capacity of civil society to collaborate and influence water governance. These changes, in turn, altered the structural context for transboundary water management, triggering adjustments at the IBWC and broadening its agenda.

But did these changes really alter the architecture of power that has historically prevailed in US–Mexico water relations? Several schools of thought may be brought to bear on the subject. One perspective that has gained currency in recent years, a view to which I partially subscribe (Mumme, Ibanez, & Till, 2012), draws on multi-governance relations theory (Hooghe & Marks, 2003; Ostrom, 2009, 2010) to argue that the new NAFTA institutions are paving the way for the emergence of a different type of transboundary water management along the border. The new management, perhaps best seen in the BECC, is grounded in a more flexible and resilient form of multilevel governance (MLG) enfolding older and more traditionally vertical modes of public administration of the sort normally associated with established federal systems. Viewed from this perspective, older established institutions like the IBWC that operate as traditional dependencies of their respective governments, even when functionally focused and enjoying significant organizational autonomy, are constrained and redirected as they become enfolded within a web of other functionally specific and relatively autonomous organizations and informal arrangements whose jurisdictions and functions overlap or touch on their own. Along the border, for example, the BECC, its institutional partner, the NADB, national border advisory groups, and new binational environmental programs fall into this category. Other coincident factors, such as political and administrative decentralization on the Mexican side of the border in the post-NAFTA period, are also contributing to changes in border area water management (Wilder & Lankao, 2006).

Another and related perspective draws on social movement and network theory with an emphasis on collaborative water governance to argue that power relations have been substantially altered in border water management. Earlier work by Francisco Lara (2000), Dan Sabet (2008), and others, documented the proliferation of non-profit advocacy networks for water management along the border. These advocacy networks are visible in the new policy arenas coming out of NAFTA. In some instances, these networks have demonstrated an ability to shape the policy agenda for transboundary water management. As a recent paper by Andrea Gerlak (2014) argues, a new MLG environment in border water governance is receptive to the informational and aggregative assets that water policy networks can mobilize as the governments try to cope with specific issues on the binational water agenda. Gerlak's recent papers (2014, Gerlak, Zamora-Arroyo, and Kahler (2013)) on the Colorado River attribute recent policy innovations in part to the sustained engagement of binational public–private research and advocacy partnerships whose knowledge, networks, and ability to leverage monetary resources have enabled the governments to tackle conservation solutions for the Colorado Delta.

These relatively recent theoretical perspectives are by no means mutually exclusive nor do they entirely contradict earlier theoretical orthodoxy on the politics of border water management. That earlier understanding of border water politics and the architecture of power in transboundary rivers management is predicated on the politics of what Robert Putnam (1988) called two-level games and grounded in an understanding of domestic politics in the two countries. From this theoretical perspective, set out in the work of policy scholars like Dean Mann (1975a, 1975b), Helen Ingram (1990), Albert Utton (1991), the diplomacy of transboundary water management is determined, first, by domestic water interest groups (we can say stakeholders) pursuing their self-interest within the effective structure of government (centralized/decentralized) and its various policy arenas on both the input and output side of the political system. These water interests are focused primarily on non-regulatory concerns related to the ownership, storage and distribution, and consumption of water. At the international level, the established treaty regime in US–Mexican relations was created to serve these respective interests as they are found in both countries, with variations in the pattern of representation and influence depending on the country. From this theoretical perspective, the prevailing pattern of politics and political power is well established and quite resistant, though not impervious, to newer and emergent social forces interested in sustainable development, ecology, and conservation, as well the achieving greater equity in the utilization of scarce water resources within and across the border region.

Each of these perspectives has a certain theoretical emphasis. MLG theorists tend to examine the institutional and political context within a policy domain. Social movement and network theorists tend to examine the way public and private policy actors interact, their communicative strategies and linkages, and the informational and material resources they bring to the policy arena. And more traditional policy analysis focuses on organized interests, established structures, rules, and procedures, differences in policy content, and the way domestic politics constrains diplomacy at the international levels.

This paper does not purport to rigorously test these various perspectives for what they say about the changing architecture of power in US–Mexican transboundary water relations but draws some insights from developments over the past decade or more in the two major river basins, parsing these through the general optics of these three perspectives. For this purpose, the same approach used in Mumme et al.'s (2012) assessment of emergent multilevel governance in US–Mexico water management is employed here, first examining Rio Grande River developments followed by a look at developments on the Colorado River.

Drought on the Rio Grande

In the post-NAFTA period (1994–2014), chronic drought has been the principal challenge for water managers on the Rio Grande (Carter, Seelke, & Shedd, 2013; Mumme, 1999, 2003). Drought conditions, punctuated by occasional extreme weather events, have prevailed for the better part of the post-NAFTA era throughout the Rio Grande River Basin, stressing water management in both countries and occasioning significant disagreement on treaty implementation between the two countries. While definitions of drought vary among climatologists, the existence of drought has a concrete operational meaning under the 1944 Water Treaty. Under the Treaty's Article 4, drought may be said to exist when Mexico fails to deliver its statutory quota of water to the USA in any given year but categorically exists when Mexico fails to deliver its treaty-mandated quota over a fixed five-year cycle (Treaty, 1944). In the past 20 years, a protracted Treaty-based drought has occurred on at least two occasions, the first in the period 1992–2005 and the second since 2007.

From a political perspective, what is interesting about the management of these two periods of Rio Grande drought is the extent to which the discussion and management of the conflict have proceeded as a full-on expression of the established management regime. To appreciate this, a brief accounting of the diplomatic development in each period is helpful.

Drought Management, 1992–2005

The Rio Grande drought crisis that developed after 1992 must be understood in terms of the treaty regime's provisions bearing on that river. Article 4 of the 1944 Water Treaty provides that Mexico supply the USA with 350,000 acre feet of water annually as averaged over a five-year cycle. It does not stipulate that Mexico is in arrears should it fail to deliver the required amount in any given year. Should Mexico be in arrears in treaty water deliveries at the end of any five-year cycle, it may request that the arrearage be carried over as debt to the next five-year cycle (Treaty, 1944, Art. 4). Left unspecified is the question of whether Mexico is obligated to pay its debt in full from the previous cycle concurrent with any debt accrued in the current cycle, or whether debt in the current cycle may be rolled over to the following cycle, with Mexico simply paying down the debt from the earlier cycle. And on this point, the two countries disagree. Also at issue is the interpretation of the Treaty's undefined reference to 'extraordinary drought' (Treaty, 1944). Article 4 provides that the Treaty's Rio Grande drought management provisions shall operate under circumstances of extraordinary drought. This means, in effect, that any water debt rollover should be triggered by a declaration of extraordinary drought, presumptively a circumstance on which both countries agree. But in the absence of any definition of the term, its effective meaning may be contested by the USA. This is no minor point as recent history has shown. Is, for example, an extraordinary drought determined by the operational fact of the amount of water delivered to the Rio Grande mainstem by Mexican tributaries, or is it determined by climatological conditions or the volume of water available in Mexico's upstream reservoirs? On that, the two countries disagree.

As a political problem, the earlier drought crisis really set in after 1997 when Mexico, in arrears on its water payments, asked for a rollover of its debt (CSIS, 2003; Vina, 2005). This was not so very controversial at the time, though such requests are always worrisome to Texas farms and municipalities downstream. When Mexico continued in arrears by 1999, however, Texas water users downstream were outraged, venting their wrath on the IBWC's US Section and the US State Department. At this point, Texas border water interests took their anger to the Texas Agricultural Commission, the Texas statehouse in Austin, Texas, and to the US Congress (Milloy, 2001, p. A14; Weiner, 2002, p. A11). Resolutions demanding Mexican compliance with the treaty's provisions were introduced in Austin and in the US House of Representatives. The issue was repeatedly raised at the level of the US–Mexico Binational Commission (cabinet-level agencies of both governments) and by 2000 in presidential summits (Wiener, 2002, p. A11). Mexico reacted by declaring an 'extraordinary drought' emergency on the River, pointing to low precipitation in the Conchos River basin. The USA responded by using satellite data to monitor levels in Mexican reservoirs, accusing Mexico of hoarding available water. Mexico's National Water Commission (CNA), in turn, rejected these accusations, noting that available supplies were insufficient to meet Mexico's own domestic demand and needs for flexibility in meeting the water needs of its upstream irrigation districts. Texas farmers indignantly argued that Mexico was using water inefficiently and had grown its irrigation districts in response to the opportunity to export crops to the USA, facilitated by NAFTA. Texas farmers would eventually (in 2004) file a NAFTA Chapter 11 lawsuit claiming monetary compensation for damages of more than $500 million USD (Associated Press, 2004; Kibel & Schutz, 2007, p. 252; Michaels, 2008).

With the dispute focused on treaty implementation, its political focus and diplomacy centered on the IBWC and the foreign ministries. By 2000, the issue had become the rarest of issues in US–Mexican water relations, one that reached the presidential level in both countries (Weiner, 2002, p. A11). Even so, the debate revolved around technical issues with each section of the IBWC coordinating with water user organizations in its country, principally irrigators, and state and federal water managers including in the USA the Texas Rio Grande Watermaster, Texas Water Development Board, and Texas Commission on Environmental Quality in Texas, and the US Department of Agriculture, US Bureau of Reclamation (BOR), and the State Department at the federal level; and in Mexico, the CNA, Chihuahua's state water commission, water users associations, and municipal utilities. Because the problem lay upstream in Mexico, there was little that US agencies could do but fulminate. Congressional resolutions vented Texas frustration and kept pressure on the executive agencies but managed little else.

At the public level, the drought crisis was at least partly the catalyst for a series of symposia and conferences focused on the sustainable management of the river. In 1999, the IBWC, the US BOR, the Western Governors Association, and Mexico's CNA sponsored a drought management workshop in El Paso, an event followed in 2000 by a federally sponsored symposium on the Rio Grande/Rio Bravo in Cd. Juarez facilitated by the natural resources and environmental agencies of the two federal governments (IBWC, 2005a). One result of the latter conference was a new Rio Grande/Rio Bravo Ecosystem Work Group established by the IBWC. The IBWC's US

Section also established a new Lower Rio Grande Citizen's Advisory Committee as part of a new public outreach program begun by then Commissioner John Bernal (IBWC, n.d.). A civil society group, the Rio Grande/Rio Bravo Coalition also convened a 'Uniting the Basin' conference in Cd. Juarez in 2000 that brought together environmental and conservation groups as well as various municipal and agricultural stakeholder groups across the basin (IBWC, 2005a). In 2001, a coalition of 22 NGOs from both countries issued a 'Binational Declaration on the Rio Conchos and the Lower Rio Grande' advancing conservation principles, ecosystem protection, attention to the hydrological cycle, the need for revised operating protocols for Mexican reservoirs, deepened binational understanding on the management of Mexican water payments, and the need to involve stakeholders in both countries in the development of drought management planning and sustainable management of the Conchos/Rio Grande River systems (Texas Center for Policy Studies, 2001).[2] In 2002, The Texas Center for Policy Studies organized a conference on the Conchos River in Chihuahua, Chihuahua and later that year, the university led US–Mexico Binational Council and the Washington, DC-based Center for Strategic and International Studies held a workshop in Austin, Texas that became the basis for a widely distributed policy report on the Rio Grande in 2003 (IBWC, 2005a). One non-profit, the Environmental Defense Fund, also focused effort on finding alternative private sector solutions, working with private irrigators along the Conchos, among other things trying to develop conservation easements with private appropriators that might reduce water demand.

Sustained discussions at the level of the IBWC and the foreign ministries, with some help from the Vicente Fox administration, led to further Mexican water releases and binational agreements in 2001 and 2002 that produced a water conservation plan designed to enable Mexico to meet its US treaty obligation (Carter et al., 2013; CSIS, 2003). The agreement, set out in IBWC Minute 308, utilized the facilities of the BECC and NADB, establishing a new Water Conservation Investment Fund (WCIF) at NADB to finance needed conservation infrastructure in Rio Conchos Irrigation District 005 in Mexico (the largest of the Conchos' irrigation districts) and along the lower Rio Grande in the USA. Minute 308 also responded to criticism of the IBWC, calling for the creation of a binational advisory committee to apprise the Commission of water management needs in the middle and lower Rio Grande basin. It promised to convene a binational conference to assess the prospects and consider ways to advance sustainable water management in the basin (IBWC, 2002).

With tensions eased somewhat, the Mexican water debt was carried over to a new cycle beginning in October 2002, the effect of which meant de facto, if temporary, acceding to the Mexican position on Article 4 (Brezosky, 2002). A binational summit on the sustainable management of the Rio Grande, attended by a wide range of stakeholders, including some nonprofit organizations, was held in November 2005 that produced a number of resolutions related to water conservation and further calls for a basin-wide advisory body (IBWC, 2005b), but little has come of this to date apart from continued implementation of the WCIF projects on the Conchos. In 2005, abetted by hurricane-driven rains that filled Rio Grande storage dams, the Mexican water debt was vacated, establishing a new five-year cycle (IBWC, United States Section, 2005a, 2005b). With that determination, the drought crisis was technically resolved.

Drought Management, 2005–2013

As some Texas irrigators were well aware, the fundamental institutional problems associated with the drought crisis that peaked after 2000 were not addressed by October 2005 when the earlier crisis ended. Essentially, the crisis was in abeyance with the help of Mother Nature, with considerably less diplomatic progress toward a satisfactory treaty-based solution. Even so, the water dispute was temporarily laid to rest. The WCIF projects went forward with the technical and financial help of BECC and NADB. The NAFTA Chapter 11-based Texas lawsuit failed (Ontario Superior Court, 2008; Shultz, 2008, pp. 12–13). Little progress was made on Minute 308 institutional reform recommendations despite the convening of the Rio Grande/ Rio Bravo Binational Summit in 2005. The Rio Bravo/Rio Grande Coalition lost momentum and temporarily ceased to exist and other NGOs have given the issue less attention since the 2005 conference. Nevertheless, the 2005–2010 cycle unfolded without any dispute.

Renewed shortages in 2011, however, alarmed Texas irrigators, triggering a new wave of accusations and protests directed at Mexico and the IBWC. Essentially, the politics of drought on the river are repeating themselves. Texas irrigation districts and municipalities are besieging state and congressional offices with resolutions demanding Mexican treaty compliance. State offices, senators, and border district congressmen are leaning on the State Department and the President. The two sections of the IBWC have met frequently since 2012 to exchange information and discuss technical aspects of the problem. Texas governor, Rick Perry, responding to local water interests, is pressing President Obama to intervene directly with Mexico's new executive, Enrique Pena Nieto (Perry, 2013). The US Ambassador, Anthony Wayne, has pressed the issue with Mexico's new administration at the cabinet level (Carter et al., 2013). Members of the Texas congressional delegation have introduced bills that would require annual reports on Mexican compliance with treaty obligations (Staples & Rubenstein, 2013) and have unsuccessfully tried to link treaty compliance on the Rio Grande to treaty compliance on the Colorado River. Mexico remains behind in its water payments and is moving forward with plans to build a series of smaller storage dams on the Conchos that Texas farmers fear will enable Mexico to unilaterally hoard treaty water and Mexican water managers argue will help store runoff to meet and better regulate treaty deliveries (Drusina, 2013; TCEQ, 2014). While there has been some progress on various conservation programs on the river, these initiatives are politically delinked from the drought crisis. As of August 2014, more than two years into the crisis and nearly four years into the current treaty delivery cycle, there is no diplomatic breakthrough (TCEQ, 2014). All parties are praying for rain.

In sum, what we see in the Rio Grande drought crisis, in both iterations, is the persistence and prevalence of traditional patterns of politics, grounded in treaty interpretation and entitlement claims, given effect through the conventional arenas of government. Dispute settlement is centered at the IBWC and the foreign ministries with the full apparatus of local, state, and national representation deployed to influence negotiations at the level of the foreign offices and presidential summitry. The NAFTA reforms are not irrelevant. As seen in the case of the WCIF, they have added additional institutional capacity to pursue certain types of solutions, financing water conservation measures in the basin. But thus far, they have proven inadequate to addressing big picture problems like recurrent drought. Civil society responses have enhanced understanding of the complex social, economic, and hydrological variables behind the drought and elevated attention to the ecological values at stake in the drought crisis but thus far remain effectively disconnected from the core politics—the demands of irrigators and municipalities channeled through state governments, federal agencies, and the national sections of the IBWC—shaping the drought dispute. Perhaps the best evidence of this is the inability of the NGO sector to convince the governments to move forward with the Minute 308 promise of establishing a binational basin-wide advisory council to work with IBWC on sustainable water management and drought mitigation practices in the basin. I shall now review developments in the other important transboundary river basin on the border, the Colorado River Basin.

Drought and Ecology on the Lower Colorado River

Persistent drought has also dogged the Colorado River Basin, particularly its lower reach. However, an important political lesson in the politics of the US–Mexico transboundary rivers is that binational relations related to the Colorado River, though governed by the same 1944 Water Treaty, are politically and institutionally separate from the management of the Rio Grande. The 1944 Treaty was designed, in effect, as two treaties in one.[3] This is something that Texas water managers dealing with the current drought have difficulty understanding as they futilely try to enlist congressional support from other US border states. These differences arise in good measure from geo-political variation between the two river basins. Whereas Texans are downstream on the Rio Grande River, on the Colorado River both Sonorans and Baja Californians share this predicament. In essence, Mexico controls most deliveries on the middle and lower reaches of the Rio Grande and the USA controls all deliveries on the Colorado. Ecological problems of binational concern are found on both rivers but loom larger politically on the Colorado River owing to the Colorado River Delta's vital role in sustaining regional biodiversity. This ecological factor inserts a less sectarian discourse based on shared national interests in recent management discussions on the Colorado.

The politics of the lower Colorado River have thus unfolded somewhat differently from the situation on the Rio Grande, a demonstration of the weak linkage between the binational management of transboundary water in the two basins. In this basin, binational advocacy networks that emerged after NAFTA have had a greater positive effect on water management in recent years. At the same time, however, the structural basis of politics at work in Lower Colorado Basin bears more than a family resemblance to what we see on the Rio Grande. This is evident if we look at two recent conservation measures on the Colorado River, the first related to the ecology of the Colorado Delta and the second involving the All-American Canal (AAC).

Delta Conservation, 2000–2012

The ecological importance of the Colorado Delta has been adequately presented elsewhere and there is no real need to review it here (Pitt, 2001; Varady, Hankins, Kaus, Young, & Merideth, 2001). Suffice it to say that persistent drought in the Colorado Basin since the late 1980s and US basin state conservation measures brokered by the BOR in the late 1990s were the catalyst for a remarkable effort by various environmental NGOs to press the IBWC for action on the conservation of the Delta ecosystem. The challenge was daunting in the face of persistent drought, the highly over-appropriated rights to water, and competing demands on the river north of the international border. In essence, there was no freshwater left in the river, not even the periodic pulse flows occasioned by flood events several times a decade that had largely sustained the Delta in the past. When state governments, particularly, California, were driven to enact more stringent water conservation measures, NGOs began to act.

Starting with a scoping conference on the Colorado Delta in 1999, a handful of national and regional environmental groups, Defenders of Wildlife, Environmental Defense, Pacific Institute, Sonoran Institute, Pronatura and others joined forces with US and Mexican universities to explore any available options for watering the Delta (Gerlak, et al., 2013; Glennon & Culp, 2002, p. 953; Pitt, 2001; Varady et al., 2001, p. 196). In 2000, they were successful in persuading IBWC to agree, in Minute 306, to support a binational task force to study the Delta problem. Unwilling to rile upstream water users with potential new claims on the river, IBWC astutely justified the initiative under the 1970 Boundary Treaty, not the 1944 Treaty. With their new-found government sponsorship, the Minute 306 Task Force set about making a case for Delta conservation, a case that included leveraging private wealth and water rights as well as public resources tied to BOR operations in the region.

By 2007, the non-profit consortium/task force formalized their collaborative effort as the Research Coordination Network (RCN) for Delta conservation. In August 2007, the governments moved to formally establish a new Core Group of lower Colorado River stakeholders building on the earlier Task Force including the major federal water agencies engaged in Colorado River management and the organized NGOs assisted by university researchers sponsored and coordinated by the IBWC. Their focus centered on how river operations and a combination of fresh and brackish water could be freed periodically to sustain the most essential features of lower Colorado River ecology and restore certain badly damaged areas (Pitt, 2001; Sonoran Institute, 2014). A vital element in this effort was the development of binational consensus on the factual basis behind ecological concerns and the fundamental requirements of the Delta ecosystem if the area was to be managed sustainably.

The opportunity to do more, faster, presented itself accidentally. A devastating earthquake hit the region in May 2010, damaging water infrastructure across the Mexicali Valley. Unable to put its treaty water to work, Mexico turned to the USA for water storage. With US upstream dams, particularly Lake Mead, severely depleted, the opportunity to temporarily store Mexican water was one the USA could hardly refuse. The RCN and the Core Group scrambled to turn this disastrous event into an even richer opportunity for US–Mexican relations, one that included Delta conservation and the inclusion of ecological concerns within the ambit of the 1944 Water Treaty.

The result, developed in a rapid succession of IBWC minutes in 2010, provided for a temporary freshwater transfer to the threatened Santa Clara Slough, this related to the recent operation of the Yuma Desalting Plant that drains concentrated saline brine to this important wetland 50 miles south of the border (IBWC, 2010a), followed by a remarkable agreement (IBWC, 2010b) that (a) established a binational Consultative Council, (b) authorized development of an operating framework for considering new conservation measures and targeted investments for that purpose, (c) linked the agreement to the 1944 Treaty, (d) for the first time referenced climate change as a drought-inducing factor, and (e) provided a basis for discussing the storage of Mexican water upstream. Minute 317 was shortly followed by Minute 318 (IBWC, 2010c) which authorized the storage of Mexican water and reinforced Minute 317 measures. As reported in an earlier article, this was an extraordinary 'breakthrough in US–Mexican water cooperation on Colorado River water supply' (Mumme et al., 2012, p. 21).

Under the authority of Minutes 317/318, the NGO/Core Group actors then proceeded to work on consolidating these gains. The operational aspects of the new agreements were calibrated at short term, allowing the parties to spend the better part of the next two years working on a more detailed agreement. In November 2012, the two governments announced a landmark agreement, Minute 319 (IBWC, 2012) that consolidated the earlier gains on two flanks to accomplish something that many thought unachievable: on the one side, agreeing to an interim five-year scheme to allocate water to the Colorado Delta ecosystem and, on the other, to establishing a binational mechanism for allocating water reductions under conditions of 'extraordinary drought'. The latter agreement is by far the more important agreement politically and involves a significant concession by Mexico in the service of managing protracted water shortages in the Colorado River Basin.

As Gerlak (2014, Gerlak et al. (2013)), Pitt (2014), Zamora (2014) and others have said, this is a remarkable achievement and one that could not have occurred were it not for the extraordinary work and political influence of non-traditional water stakeholders (NGOs, universities and research institutes) working in tandem with traditional stakeholders in the Colorado (particularly Lower Colorado) River Basin. These non-traditional stakeholders offered a range of resources that augmented those of government agencies and irrigation districts, including scientific expertise, research and policy networking, and the ability to utilize private water markets to facilitate consideration of binational solutions to the Delta conservation problem Gerlak et al., 2013; Mumme et al., 2012). They drew on NAFTA-related reforms at the IBWC, its new citizen advisory boards, groups whose efforts were at least partly facilitated by strengthened La Paz Program initiatives for environmental protection on the border. Though they have not yet utilized the resources of BECC and NADB, this is certainly in the mix of infrastructure development options available should the need arise. The new multilevel governance context was helpful here. Coupled to the Mexicali Valley earthquake disaster and the opportunity it provided for innovative thinking on the river, these citizen-based organizations were able to make a compelling case that it would be in the mutual self-interest of Mexican and US water users to make needed sovereign concessions—the USA providing Mexico needed water storage and releasing water to the Delta; Mexico accepting a drought water rationing formula modestly favoring the USA; both countries recognizing ecological preservation as a beneficial use of Colorado River water—that would safeguard their water supplies under conditions of prolonged scarcity down the line. This was nothing short of a dramatic shift in conventional thinking on the law of the river that moves its management in a more sustainable direction and binationally cooperative direction. However, a cautionary note should be appended here, as witnessed in the political dynamics of the AAC dispute during this same period.

The AAC Controversy

At roughly the same time that Colorado River water conservation needs north of the border were driving collaborative efforts to save the Colorado Delta,[4] the same dynamics contributed to a binational dispute over the long proposed lining of the AAC. The AAC, completed in 1940 prior to the signing of the 1944 Water Treaty, was dug to provide the giant Imperial Irrigation District (IID) with a delivery system for Colorado River water independent of an older canal, the Alamo Canal, that crossed the Mexicali Valley before draining to the USA (Mumme & Lybecker, 2004). The AAC runs roughly 80 miles from the intake at Imperial Dam, of which nearly 40 miles parallel the international boundary at short distance. The fact the canal was unlined meant considerable seepage southward to Mexico, water that Mexican farmers, wildlife, and waterfowl came to depend upon (Cortez Lara, 2006; Zamora-Arroyo, 2006).

Efforts to reclaim this seepage dates as far back as the early 1980s as pressure mounted on the IID to use water more efficiently so that surplus could be shared with other Southern California water claimants. In 1988, the San Luis Rey Indian Water Rights Settlement Act gave the Interior Department and its BOR the authority to devise a project to recapture AAC seepage and by 1994, BOR settled on a plan to build a new concrete lined canal running parallel to the older ditch. Implementation was delayed, however, due to financing constraints—the 1988 Act mandated that the project's beneficiaries pay its costs. When California was forced to pare down its draw on the Colorado River in 1999, it pushed through the California Quantification Settlement Agreement in 2003, a statewide conservation measure that upped the ante on lining the AAC. A deal with the San Diego County Water Authority was struck by which San Diego would foot the cost in exchange for the right to buy IID water for the next century. At that point, the BOR went forward with the project.

The lining project had drawn Mexico's objections at the level of the IBWC and Mexico's Foreign Ministry for years and Mexico continued to quietly question the international legitimacy of the project. While various arguments could be made against the lining under international law, Mexico did not forcefully advance its objections, unwilling to antagonize the USA at a time the NAFTA agreement was being negotiated and implemented. When BOR announced that it would proceed with the project in 2003, local interests in the Mexicali Valley were incensed, as were many environmentalists. A binational coalition quickly formed to oppose the project in US courts that joined agricultural, business, municipal, and environmental groups throughout the region.

In 2006, three NGOs filed suit, the Consejo de Desarrollo Economico de Mexicali (CDEM), a Mexicali business group, Citizens United for Resources and the Environment, a US regional environmental group, and Desert Citizens Against Pollution, another regional environmental group championing environmental justice. The City of Calexico also lent support (Ries, 2008, p. 504). Other environmental organizations working on the Colorado Delta problem were quietly supportive, but with the possible exception of Defenders of Wildlife, did not join the cause.[5] The plaintiffs based their lawsuit on several sets of claims, arguing that lining the canal violated certain Mexicali Valley water rights, would harm the Mexicali economy, violated several federal statutes including the US National Environmental Policy Act, the US Administrative Procedures Act, the US Endangered Species Act, and the Migratory Bird Treaty Act, and violated terms in the San Luis Rey Indian Reservation Water Rights Settlement Act (Ries, 2008, pp. 505–507).

The US District Court in Las Vegas, Nevada denied almost all of these claims, finding that the Mexican plaintiffs had no standing to sue, that the laws in question lacked extraterritorial application, and that the US environmental groups lacked 'associational standing'. After an amended filing, the court accepted two of the plaintiffs' claims; (1) that BOR had failed to file a Supplemental Environmental Impact Statement (SEIS) after learning new information concerning the project's impact on the previously unknown Andrade Mesa Wetlands and (2) that BOR had violated the Endangered Species Act. When the plaintiffs demanded summary judgment of the two remaining claims, they were denied (Ries, 2008, p. 512).

The plaintiffs then appealed to the US 9th Circuit for an injunction that would halt the project pending preparation of further appeals. At the 9th Circuit, US attorneys claimed 'sovereign immunity', arguing that the seepage was US water governed by the 1944 Treaty and that CDEM's arguments were thus 'non-justiciable political questions' requiring a diplomatic rather than judicial resolution. The City of Calexico intervened in the suit in support of the plaintiffs and made the case that NAFTA effects had not been taken into account in the 1994 BOR-FEIS such that the adverse effects of lining the canal had not been properly understood. The 9th Circuit granted an injunction temporarily halting the project (Ries, 2008, p. 514). The plaintiffs gained a ray of hope when the court admonished BOR for failing to take new information into account.

At this point, however, the project's supporters placed their faith in a parallel effort to influence the US Congress. Southern California's congressional delegation successfully attached a bill to the Tax Relief and Health Care Act of 2006 ordering the Secretary of the Interior to build the project and specifying that the 1944 Treaty was, 'the exclusive authority for identifying, considering, analyzing, or addressing impacts occurring outside the boundary of the United States of works constructed, acquired, or used within the territorial limits of the United States (H.R. 6111, 2006). The 9th Circuit's injunction, and its specific judicial authority in the case, was effectively nullified and the lawsuit died (Ries, 2008, p. 515).

In sum, what we see in the AAC case is the classic resort to sovereignty and traditional water management politics north of the border asserting water allocation claims under the drawing on the 1944 Treaty. In this case, established water interests employed the tools of legislative, executive, and judicial power at the federal and state levels to prevail over a well-organized binational coalition of mostly non-traditional, non-profit actors on claims of water entitlement, ecological sustainability, and binational equity. It is interesting to note that the Delta Coalition, their resources and networks as arrayed in 2006, appear to have had marginal effect on the AAC dispute. In fact, the major non-profit players, with one notable exception, Defenders of Wildlife, were quite reluctant to join the case as plaintiffs—even Defenders declined to do so. It is evident that as effective as the NGOs were on the Colorado Delta issue, they were mostly unwilling to jeopardize their fragile and hard won political capital in one arena on another, particularly one that was as controversial and which had the support of the most powerful water interests in Southern California. It is interesting to speculate what might have happened had the Mexican Foreign Ministry, Secretaria de Relaciones Exteriores (SRE) sought to support the binational coalition by challenging the AAC lining with appeal to the 1944 Water Treaty. We do not know but can well suppose that technical experts at SRE and CNA may have believed that they had a weaker case in international law or thought the consequences of challenging the USA from anything less than a strong hydrological and legal position would invite sanctions or a hardened position on other bilateral issues, including in the case of a reoccurrence of drought on the Rio Grande. What we do know is that the Mexican CDEM was only able to get a soft statement of concern out of SRE during this period and that its concerns were not really pushed by the Mexican Section of the IBWC, Comision de Limites y Aguas.

Theoretical Perspectives and Post-NAFTA Water Security on the US–Mexico Border

The recent history of disputes and cooperation on the two major transboundary rivers suggests that the basic structure of politics and power, the architecture of power, affecting water security along the border has not changed very much. This traditional system of water security is dominated by sovereign defense of treaty-established water endowments resting upon a power structure formed largely of vested surface freshwater rights (now even in Mexico), water development interests, urban and agricultural water districts, state water agencies (more powerful in the USA but growing in political importance in Mexico), and federal water, development, and financial agencies that make up the water policy sector. In contrast, environmental and natural resources agencies have limited voice in transboundary water policy.

While various analysts are impressed by post-NAFTA changes in border institutions that have generated attention to water conservation, environmental protection, and sustainability concerns, it is also easy to overstate their effects. When we look at these disputes, we do see some remarkable accomplishments, as witnessed recently on the Colorado River, that advance certain aspects of water security, certainly biodiversity protection and drought management that are attributable to the emergence of unprecedented collaborative efforts in a particular issue-area that have changed the architecture of power at least slightly. When we look at the AAC dispute, it is also evident that water governance reforms in one policy realm do not necessarily translate to another and that the established structure of power in border water governance remains well embedded.

If we look at the Rio Grande drought disputes, where BECC and NADB have played a modest role in financing needed upstream conservation measures, we have yet to see much institutional development. Over a dozen years have passed since the governments promised to create a basin-wide advisory committee to help IBWC better manage shortages and nothing much has come of it. Sovereign interest prevails and we see this in the dynamics of these disputes which pursue a very familiar political path wherein local and state interests bring political pressure to bear on both the executive and legislative side of US federal government with the hope that the president can persuade his Mexican counterpart to be more responsive to US concerns. Mexico's president, of course, has limited domestic incentive to comply, but the long-standing asymmetry in US–Mexican relations means he cannot ignore a strong appeal by the US executive. And so, we see a familiar political theater play out today much as it did a decade ago.

More recent theoretical perspectives on border water governance should be seen in this light. None of the theories mentioned above are really mutually exclusive. They point to different facets of a complex reality affecting water availability along the border.

The older theories of water politics drawing on political pluralism or, in the Mexican case, corporatism or presidentialism, should still be taken seriously because they describe a deeply embedded and treaty-reinforced system of national and binational water governance that still operates much as it did 70 years ago. New theoretical perspectives capture what is changing in US–Mexican water relations; new institutions, emergence of collaborative social networks, additive policy emphasis on sustainability and conservation, environmental protection and public health, and greater openness to public participation, or at least better public relations in water development. But these reforms have just begun to make a modest dint in US–Mexican transboundary water governance and this can be seen in the treaty regime itself, which has been extraordinarily resistive to political change—as we would expect of a constitutional treaty arrangement.

The problem, of course, evident in each of the cases outlined above (all of which can be described as responses to prolonged water scarcity, or drought), is that many of the embedded assumptions associated with the treaty allocation of freshwater on the transboundary rivers are changing with the climate and the enormous demands now placed on these limited resources. There is a need for greater flexibility in the treaty regime and this must be accomplished politically. The open question is whether non-traditional stakeholders in sustainable water management will gain a more influential role in transboundary water management as the problem of water scarcity, and water insecurity, deepens along the US–Mexico border.

Disclosure Statement

No potential conflict of interest was reported by the author.

Notes

1   This is seen in the 1944 Treaty's Article 3, which specifies priorities for the beneficial use of Treaty water. The availability of water for domestic and municipal use, agriculture, hydroelectric power, and other industrial uses, water uses endowed with clear and tangible economic value and the core values to be gained by formal allocation, is given priority over other potential beneficial uses whose values are arguably less tangible—true even for navigation which at the time of the treaty was understood as impracticable on the treaty rivers and included so as to extend the reach of the U.S. federal commerce clause to the administration of the rivers. Allocation also trumps the Treaty's Article 3 border sanitation provisions which are clearly subsidiary to the Treaty's main purpose. The Treaty is also acknowledged to supersede the La Paz Agreement's pollution protection provisions as those bear on treaty water management and prevails over other binational executive agreements on natural resources management should conflicts arise.

2 The effort was spearheaded by Texas Center for Policy Studies. Signatory NGOs included World Wildlife Fund (USA and Mexico), Pronatura Noreste, Bioconservacion, Environmental Defense, Texas Center for Policy Studies, Alliance for the Rio Grande Heritage, Southwest Environmental Center, and Rio Grande/Rio Bravo Coalition among others (Texas Center for Policy Studies, 2001, p. 6).

3 Some might say three treaties insofar as the 1944 Treaty addresses the Tijuana River as well.

4 Active binational discussion of the All American Canal question dates to before 1983 and intensified in earnest after 2001 with the adoption of strict water conservation rules north of the border. Active discussion of the Delta began, as noted above, in 1999 and intensified after 2007, after the AAC issue had been decided in U.S. courts.

5 Personal communication with William Snape, Defenders of Wildlife, and further discussion with Melissa McKeith, an attorney representing CURE, 2006.

References

Associated Press. (2004, August 27). Farmers: Mexico water debt due: Valley group files NAFTA claim for $500 million for losses.

Brezosky, L. (2002, October 1). *Mexico's water deadline expected to pass without water release.* Associated Press.

Carter, N. T., Seelke, C. R., & Shedd, D. T. (2013, November 19). *U.S.–Mexico water sharing: Background and recent developments.* Washington, DC: Congressional Research Service, CRS Report 7–5700.

Cortez Lara, A. (2006). Opposing approaches to managing share water resources: The lining of the All-American Canal and the Valley of Mexicali—Static market equilibrium or Nash equilibrium. In V. S. Mungia (Ed.), *Lining the All-American Canal: Competition or cooperation for water across the U.S.–Mexico border* (pp. 197–212). San Diego, CA: SCERP Monograph, No. 13, San Diego State University Press.

CSIS. (2003). *U.S.–Mexico transboundary water management: The case of the Rio Grande/Rio Bravo.* Washington, DC: US-Mexico Binational Council—CSIS, Instituto Tecnologico Autonomo de Mexico, University of Texas at Austin.

Drusina, E. (2013, April 5). Letter by the US Section, IBWC, commissioner to Texas congressmen, honorable Henry Cueller, Honorable Pete Gallego, Honorable Ruben Hinojosa, Honorable Filemon Vela.

Gerlak, A. (2014). Resistance and reform: Transboundary water governance in the Colorado Delta (Unpublished paper).

Gerlak, A., Zamora-Arroyo, F., & Kahler, H. (2013, April). A Delta in repair: Restoration, binational cooperation, and the future of the Colorado River Delta. *Environment.*

Glennon, R., & Culp, P. (2002). The last green lagoon: How and why the Bush administration should save the Colorado River delta. *Ecology Law Quarterly, 28*(4), 903–992.

Hooghe, L., & Marks, G. (2003). Unraveling the central state, but how? Types of multi-level governance. *American Political Science Review, 97*(2), 233–243.

IBWC. (2002, June 28). *Minute No. 308, United States allocation of Rio Grande waters during the last year of the current cycle.* Ciudad Juarez, Chihuahua.

IBWC. (2005a). Binational Rio Grande summit: Cooperation for a better future. Background document. El Paso: IBWC. Conference held November 17–18, Reynosa, Tamaulipas and McAllen, Texas. Retrieved from http://www.ibwc.gov/Organization/rg_summit.html

IBWC. (2005b). Binational Rio Grande summit: Cooperation for a better future. Recommendations of the work group on environment and water quality. El Paso: IBWC. Conference held November 17–18, Reynosa, Tamaulipas and McAllen, Texas. Retrieved from http://www.ibwc.gov/Organization/rg_summit.html

IBWC. (2010a, April 16). *Minute No. 316, utilization of the Wellton-Mohawk bypass drain and necessary infrastructure in the United States for the conveyance of water by Mexico and non-governmental organizations of both countries to the Santa Clara Wetland during the Yuma Desalting Plant pilot run.* Yuma, AZ.

IBWC. (2010b, June 17). *Minute No. 317, Conceptual framework for U.S.–Mexico discussions on Colorado River cooperative actions.* Ciudad Juarez.

IBWC. (2010c, December 17). *Minute No. 318, Adjustment of delivery schedules for water allocation to Mexico for the years 2010 through 2013 as a result of infrastructure damage in Irrigation District 014, Rio Colorado, caused by the April 2010 earthquake in the Mexicali Valley.* El Paso, TX.

IBWC. (2012, November 20). *Minute 319, interim international cooperative measures in the Colorado River Basin through 2017 and extension of Minute 318 Cooperative measures to address the continued effects of the April 2010 earthquake in the Mexicali Valley.* Coronado, CA.

IBWC. (n.d.). Lower Rio Grande Forum meetings. Retrieved from http://www.ibwc.gov/Citizens_Forums/CF_Lower_RG.html.

IBWC, United States Section. (2005a). *Mexico pays Rio Grande water debt.* El Paso: IBWC, US Section. Retrieved from http://www.ibwc.state.gov/PAO/CURPRESS/2005/WaterDebtPaidFinal.pdf

IBWC, United States Section. (2005b). *USIBWC commissioner announces resolution of Mexico's Rio Grande water debt.* El Paso: IBWC, US Section. Retrieved from http://www.ibwc.state.gov/PAO/CURPRESS/2005/WaterDelFinalWeb.pdf

Ingram, H. (1990). *Water politics: Continuity and change.* Albuquerque: University of New Mexico Press.

Kibel, P. S., & Schutz, J. (2007). Rio Grande designs: Texans' NAFTA water claim against Mexico. *Berkeley Journal of International Law, 25,* 101–143.

Lara, F. (2000). Transboundary networks for environmental management in the San Diego-Tijuana region. In L. Herzog (Eds.), *Shared Space: Rethinking the U.S.–Mexico Border Environment* (pp. 155–184). San Diego, CA: Center for US–Mexican Studies, University of California, San Diego.

Mann, D. (1975a). Conflict and coalition: Political variables underlying wáter resource development in the upper Colorado River Basin. *Natural Resources Journal, 15*(1), 141–170.

Mann, D. (1975b, January). Politics in the United States and the salinity problem of the Colorado River. *Natural Resources Journal, 15*(1), 113–128.

Michaels, D. (2008, February 18). South Texans take water fight with Mexico to Canada court. Dallas Morning News.

Milloy, R. E. (2001, September 18). A rift over Rio Grande water rights. *New York Times,* p. A14.

Mumme, S. (1999). Managing acute water scarcity on the U.S.–Mexico border: Institutional issues raised by the 1990's drought. *Natural Resources Journal, 39,* 149–166.

Mumme, S. (2003). Revising the 1944 Water Treaty: Reflections on the Rio Grande drought crises and other matters. *Journal of the Southwest, 45*(4), 649–670.

Mumme, S., & Ibanez, O. (2013). Power and cooperation in U.S.–Mexico water management since NAFTA. In P. Gilles, H. Koff, C. Maganda, & C. Schultz (Eds.), *Theorizing borders through analysis of power relationships* (pp. 151–176). Brussels: P.I.E. Peter Lang.

Mumme, S., Ibanez, O., & Till, S. T. (2012). Multilevel governance of water on the U.S.–Mexico border. *Regions & Cohesion, 2*(2), 6–29. doi:10.3167/reco.2012.020202

Mumme, S., & Lybecker, D. (2004). El Canal Todo Americano: perspectivas de la posibilidad de alcanzar un acuerdo binacional. In V. S. Mungia, coordinador, *El revestimiento del Canal Todo Americano* (pp. 217–246). Tijuana: Colegio de la Frontera Norte.

Ontario Superior Court. (2008). Bayview irrigation district no. 11, et. al. v. United Mexican States. COURT FILE NO.: 07-CV-340139-PD2. May 5. Retrieved from http://italaw.com/sites/default/files/case-documents/ita0078_0.pdf

Ostrom, E. (2009). A general framework for analyzing sustainability of social-ecological systems. *Science, 325,* 419–422. doi:10.1126/science.1172133

Ostrom, E. (2010). Polycentric systems for coping with collective action and global environmental change. *Global Environmental Change, 20,* 550–557. doi:10.1016/j.gloenvcha.2010.07.004

Perry, R. (2013, April 9). Texas Governor Rick Perry letter to Honorable Barack Obama, President of the United States.

Pitt, J. (2001). Can we restore the Colorado Delta? *Journal of Arid Environments, 49*(1), 211–220.

Pitt, J. (2014, June 1–3). Presentation to uncommon dialogue: Workshop on US–Mexico trans-boundary water issues at Stanford Law School.

Putnam, Robert. (1988). Diplomacy and domestic politics: The logic of two-level games. *International Organization, 42*(3), 427–460. doi:10.1017/S0020818300027697

Ries, N. (2008). The (almost) All-American Canal: Consejo de Desarrollo Economico de Mexicali v. United States and the pursuit of environmental justice in transboundary resources management. *Ecology Law Quarterly, 35,* 491–528.

Sabet, D. (2008). *Nonprofits and their networks: Cleaning the waters along Mexico's northern border.* Tucson: University of Arizona Press.

Shultz, J. (2008, September 1–4). *Applying California water rights takings jurisprudence to international water rights expropriation cases.* Paper presented at the 13th IWRA International Water Congress, Montepelier, France. Retrieved from http://www.iwra.org/congress/2008/index.php?page = roceedings&abstract_id = 156

Sonoran Institute. (2014). *Colorado River Delta program: Restore, renew, reconnect.* Tucson: Sonoran Institute. Retrieved from http://www.sonoraninstitute.org/component/docman/doc_details/1551-colorado-river-delta-program-summary-10152012.html?Itemid = 3

Staples, T., & Rubenstein, C. (2013, June 8). Addressing Mexico's water deficit to the United States. *Texas Commission on Environmental Quality.* Austin: TCEQ. Retrieved from http://www.texasagriculture.gov/Portals/0/forms/COMM/Water%20Debt.pdf

Texas Center for Policy Studies. (2001). *Binational declaration.* Austin, TX: Author.

Texas Commission on Environmental Quality. (2014, August 16). *Water shortage issue related to Mexican water deficit.* Austin: TCEQ. Retrieved from http://www.tceq.texas.gov/border/water-deficit.html/#issue

Treaty regarding utilization of the waters of Colorado and Tijuana Rivers and of the Rio Grande. (1944, February 3). U.S.–Mexico, 59 Stat. 1219.

Utton, A. (1991). International waters: International streams and lakes; Canadian international waters; Mexican international waters. In R. Beck (Ed.), *Waters and water rights* (pp. 1–128). Charlottesville: The Michie Company.

Varady, R. G., Hankins, K. B., Kaus, A., Young, E., & Merideth, R. (2001). To the sea of Cortes: Nature, water, culture, and livelihood in the lower Colorado River basin and delta—an overview of issues, policies, and approaches to environmental restoration. *Journal of Arid Environments, 49*(1), 195–211. doi:10.1006/jare.2001.0842

Vina, S. (2005). *The United States—Mexico dispute over the waters of the lower Rio Grande River.* Washington, DC: Congressional Research Service, RS22085, March 21.

Wiener, Tim. (2002, May 24). Water crisis grows into test of U.S.–Mexico relations. *New York Times*, p. A11.

Wilder, M., & Romero Lankao, P. (2006). Paradoxes of decentralization: Water reform and social implications in Mexico. *World Development, 34*(11), 1977–1995.

Zamora, F. (2014, June 1–3). Presentation to uncommon dialogue: Workshop on US–Mexico trans-boundary water issues at Stanford Law School.

Zamora-Arroyo, F. (2006). Looking beyond the border: Environmental consequences of the All-American Canal project in Mexico and potential binational solutions. In V. S. Mungia (Ed.), *Lining the All-American Canal: Competition or cooperation for water across the U.S.–Mexico Border* (pp. 21–58). San Diego, CA: SCERP Monograph, No. 13, San Diego State University Press.

Many Faces of Security: Discursive Framing in Cross-border Natural Resource Governance in the Mekong River Commission

ANDREA K. GERLAK & FARHAD MUKHTAROV

ABSTRACT   *In the past decade, security has emerged as a new discourse in water governance beyond transboundary conflicts and cooperation. This paper will examine how security is framed in the context of international river basin organizations (RBOs), key regional organizations in transboundary water governance operating in many international river basins around the world. As an example of cross-border governance, RBOs can promote joint cooperation and information sharing, and serve as a form to bring together diverse stakeholders. This paper focuses on the discursive construction of 'security' in a particular context of cross-border river basin governance in the Mekong River Basin. We ask: How is security framed in the discourse of RBOs? We examine how diverse actors frame security in the context of RBOs and at various scales and around certain management actions in a case study of the Mekong River Commission, a well-established RBO. Attention will be paid to the links between water security, food security, and energy security in the broader water and development discourse. We analyze what the findings mean for cross-border governance more broadly.*

Introduction

The concept of security has expanded to many areas of social sciences in the last several decades, such as environmental governance, public policy, and international development. Because security as a concept is 'the product of an historical, cultural, and deeply political legacy'

(Williams, 2007, p. 1), it is most telling to discuss it in the context of a particular resource, issue, or region where securitization happens. In different context, security may refer to human security, environmental resource security, and more particularly, water, energy, or food security (Fischhendler & Katz, 2013). Defining 'security' as divorced from the context is unnecessary for this paper since its focus is on the discursive construction of 'security' in a particular context. Our ontological assumptions therefore claim that the term 'security' acquires meanings only in a particular context of its use, which is the primary subject of our attention.

Securitization, in turn, refers to introducing security jargon and assumed rules of decision-making to areas that have previously been accepted as 'normal politics' and within the remit of public debate and decision-making (Buzan, Waevar, & de Wilde, 1998; Fischhendler & Katz, 2013). This has further contributed to a shift away from a narrow definition of security as linked to military conflicts toward a broader definition of security to include the threats and vulnerabilities that can arise in many different areas. According to Buzan et al. (1998), in the environmental sector, security presumes crossing certain thresholds that signify the limits for survival of species, destruction of their habitat, and economic and social affluence that are linked to the use of natural resources. With the regional overtones, this becomes an issue of security in certain contexts, such as the European Union (EU), North American Free Trade Agreement (NAFTA), or Association of Southeast Asian Nations (ASEAN). In this paper, we are concerned with securitization of various natural resources and issues in the region of Southeast Asia as defined by the Mekong River Basin.

In the context of transboundary waters, some scholars have called attention to framing and discourse in order to highlight or subvert particular dynamics of state relations (Zeitoun & Warner, 2006, p. 448), or to reinforce particular problems like water scarcity (Mirumachi, 2010), or climate change (Gerlak & Schmeier, 2014). In water governance, actors may use certain frames to target a particular audience to help boost their legitimacy or enroll new actors in their coalitions (Mukhtarov & Gerlak, 2013). Scale may be invoked in discourse to frame problems and solutions, incorporate or exclude actors, and challenge, heighten influence or power, or legitimize power asymmetries (Lebel, Garden, & Imamura, 2005; Mansfield, 2005). Actors may adopt multiple-scale discourse to call attention to an issue, and then may shift or change scales to expand or limit the suite of available solutions (Harrison, 2006).

There is little research exploring how water, energy, and food security language is constructed and conceptualized as opposed to other concepts and models in the field of environmental management. The aim of this study is to address this gap by examining how resource security is framed, and what proposed policy strategies and management actions, at what scale, are proposed to address security concerns and threats. We examine the security discourse in the context of cross-border areas by studying the official discourse of the Mekong River Commission (MRC). Building from a long-standing interest in international relations in how discourse articulated by political actors produces policy actions (Milliken, 1999), we draw on the term discourse to analyze how the language and practices of security are problematized and produced.

An understanding of the use of securitization discourse in an international river basin organization (RBO) can help shed light on the differing meanings attributed to the discourse and motivation for its adoption and use. It is important to be able to understand what resources are discussed in security language, and how (Fischhendler & Katz, 2013). Discourse and framing can ultimately impact water governance and the nature of interstate relations and cooperation. It is evidenced that the increasing securitization of water resources may result in lessened cooperation between the countries, who perceive this as the matter of 'high politics' and are less likely to compromise (Mukhtarov & Cherp, 2014). At the same time, with the increasing popularity of 'water security' and 'energy security' discourses globally, it is important to observe whether these find any resonance at the regional scale in one of the world's major transboundary river regimes. Our case study can report on the scope and extent of securitization in cross-border water governance in the Mekong area and beyond and contrast the results of the actual securitization with some of the expectations based on the literature. Our paper, to the best of our knowledge, is among the first ones to examine securitization in RBOs through the study of their documents, which may be an interesting research avenue to pursue for the future researchers.

We first examine how 'security' is framed globally and how the concept has evolved over the past two decades. Then we examine the role of RBOs in transboundary water governance. We present an overview of the MRC and a discussion of emerging pressures in the basin. Next, we present the findings of our content analysis of the MRC official discourse over the past years. Our discussion explores how security and scale are framed by the MRC, including possible implications for cross-border water governance. We discuss how our findings correspond to the expectations of securitization of various resources based on the security literature broadly and environmental security more narrowly, and to those linked to the regional focus of our study. Our conclusion summarizes our findings and outlines potential directions for future research.

Security and the Environment

The use of security in the context of a geographical region and an RBO may take several forms. *Human security*, the concept launched in 1994 by the United Nations Development Programme, combines the political-activist and academic agendas and is often seen as an inclusive category for other conceptualizations of security (Buzan & Hansen, 2009). Proponents of human security argue that the 'logic of security' needs to be broadened to include 'universal concerns' of prevention of conflicts, and be of global nature in order to soften the North–South divides, eradicate poverty, and underdevelopment. However, critics argue that if human security comes to mean so many things, it may lose meaning and turn into a solely political token (Paris, 2001). Others raise concerns about the value of mixing up the security agenda with what has traditionally been seen as a human rights agenda (Buzan, 2004). Human security has been widely adopted in international development and humanitarian foreign policies, perhaps as a corollary to its conceptual vagueness, intuitive appeal, and moral attractiveness (Buzan & Hansen, 2009).

Another related concept is *environmental security*. This discourse has been manifest since the 1972 United Nations Conference on the Human Environment. The environmental security discourse is bi-polar; one side is dominated by environmental science and the other by political agendas. The agenda of environmental security includes: (i) state and public awareness about the environmental issues; (ii) the acceptance of political responsibility for dealing with these issues; and (iii) the political management questions that rise: problems of international cooperation and institutionalization—such as regime formation and diffusion of norms in transnational environmental governance (Buzan et al., 1998). The Toronto school of environmental security has emphasized the scarcity of resources as the major source of insecurity (e.g. Homer-Dixon, 2000). At the same time, the agenda promoted by environmental security is broad: from disruption of ecosystems and energy issues to food security and civil rights. NATO's program on environmental security illustrates the increased institutionalization of environmental security, although, as with human security, there are opponents of securitization of environmental issues (Fischhendler & Katz, 2013).

In the context of human security, environmental security has been linked to poverty reduction concerns (Adeel, 2012; Bogardi et al., 2012), and the human right to water (Gerlak & Wilder, 2012). Environmental security can be seen at the heart of struggles for economic growth, poverty reduction, and sustainable development (Grey & Sadoff, 2007, p. 547). It may also be seen as an extension of key aspects of integrated water resources management (IWRM), including linkages between land-use change and hydrological systems, between ecosystems and human health, and between political and scientific aspects of water management (Bakker, 2012). Reports and research on environmental security commonly point to calamities caused by a multitude of factors, including floods, droughts, natural disasters, pollution, epidemics, and climate-related events (Brzoska, 2009; UN-Water, 2009). Water security is also often considered within the security triad of environment–energy–food (Wouters, Vinogradov, & Magis, 2009, pp. 98– 99). Given its growing prominence, some authors consider the water security concept to take on IWRM as the major discourse in water management (e.g. Gerlak & Mukhtarov, 2015).

The third related concept is that of *food security*. Food security usually refers to hunger that may trigger various conflicts (Blakeney, 2009). It is often discussed in the context of development and agriculture, and lately also in the context of energy security as linked to the production of biofuels and the possible hikes in food prices as a consequence (Asveld, van Est, & Stemerding, 2011). There is much uncertainty with regard to how important food security can become on an international political agenda. In addition to the discourse of ensuring sufficient food for one's population, there is also a discourse of food security as ensuring that the bulk of the food is homegrown. According to Falkenmark and Lundquist (1998) and Falkenmark (2013), there is a clear link between the water endowment of a country and its self-sufficiency in terms of food supplies. The 'virtual water' thesis has been suggested by Allan (1995), who advocated the international trade and transfer of food rather than water resources. However, the political concerns in terms of food supply interruptions as linked to political factors typically prevent the idea of 'virtual water' from becoming predominant (Biswas, 1999). Overall, the food security remains a potential game-changer in natural resources management,

> (a)s the sudden upsurge of concern about food security in 2007–2008 showed, when the diversion of agricultural production to biofuels helped, along with high oil prices, to jack up the price of many basic foods, they are capable of changing the game quickly and radically. (Buzan & Hansen, 2009, p. 268)

Another concept that has gained in importance in recent years is *energy security*, an umbrella term that is often used to describe many concerns such as a potential threat to standards of living, economic growth, or political power and stability that may materialize from lack of access to energy-generating resources (Hughes, 2009). According to Cherp and Jewell (2011), there are two notable schools of energy security: one which has to deal with the issue of geo-politics in terms of supplies of energy sources and questions such as 'peak oil'; and the other of global governance with the focus on institution-building and regime formation, including international markets for distribution of energy. They also emphasize the national-level context within which energy security challenges emerge, and within this context, the major attention would be paid to the physical, political, and economic issues (Cherp & Jewell, 2011, p. 211).

Finally, the most recently risen concept of *water security* attracts the attention of analysts and practitioners in the field of public policy and human geography. In the 1990s, water security was typically linked to human security issues like economic security, environmental security, food security, or military security (Cook & Bakker, 2012, p. 97; Liotta, 2002). The focus was on human needs. But the concept of water security has shifted to include an ecosystem or sustainability element. This includes a focus on guaranteeing the functioning of the biosphere as the basis for human well-being and existence (Lopez-Gunn, De Stefano, & Llamas, 2012, p. 91). According to Grey and Sadoff (2007, p. 545) water security is focused on 'the availability of an acceptable quantity and quality of water for health, livelihoods, ecosystems and production, coupled with an acceptable level of water-related risks to people, environments and economies'.

All in all, it is clear that 'security' is a highly malleable term that is open to multiple interpretations and an evolving conceptual shaping. The concept takes on distinct meanings and interpretations based on the audience, context, and disciplinary approach. In the global water context, security as a discourse has been on the rise since the new Millennium; however, it is an interesting question whether this emphasis on security and the links it brings to food and energy security have been also visible at the regional level of water governance, and the subsequent policy strategies and management actions that accompany it. In order to examine this relationship, we will look at the regional level discourse on RBOs and the treatment of security in this arena.

RBOs and Cross-border Water Governance

To overcome complex collective action problems in transboundary river basins, riparian states may create international RBOs, under the framework of international water treaties, to jointly govern shared resources at the basin scale (Wolf, 2007). RBOs can promote joint cooperation and information sharing, and serve as a forum to bring together diverse stakeholders (Marty, 2001; Mukhtarov, 2009). RBOs also play an important role in gathering, analyzing, and disseminating information (Burton & Molden, 2005), and supporting IWRM (Molle & Wester, 2009). In the developing world, RBOs also often play an important role in capacity building within their member-states by providing technical and management knowledge to water resources managers, often facilitated by international donors (e.g. Van Harten, 2002). From an international perspective, RBOs are on the rise today, promoted by a variety of intergovernmental and non-governmental organizations (Conca, Wu, & Ciqi, 2006; Gerlak & Grant, 2009; Mukhtarov, 2013; Schmeier, Gerlak, & Blumstein, 2015). RBOs have been advanced as part of the broader neo-liberal discourse and the emphasis on decentralization (Warner, 2006).

Recent research examining global water discourse and governance highlights the role of transnational policy entrepreneurs, particularly global knowledge networks, in producing and maintaining the global discourse of RBOs through the development of knowledge, development assistance projects, global water meetings, and publications (Mukhtarov & Gerlak, 2013). For example, for networks like the Global Water Partnership and the World Water Council, institutions at the river basin level are thought to help IWRM, an approach to water management which aims at holistic and multi-level governance of water and related resources (Mukhtarov, 2007).

A growing number of actors have been linking RBOs and river basin governance to water security (e.g. Leb & Wouters, 2013; Tarlock & Wouters, 2009; UN-Water, 2013). The link made between RBOs and water security is twofold. On the one hand, RBOs have been an instrument to render the transboundary politics technical through cooperation agreements, joint monitoring, and procedures such as early warning and routine information exchanges. By engaging with these technical tasks, trust is built and compromise among riparians is nurtured. For example, Zeitoun, Eid Sabbagh, Talhami, and Dajani (2013) argues that water security can be achieved only through fostering collaboration between concerned nation states. This is the strand of the literature that views transboundary RBOs as a gateway to water security. Researchers highlight the importance of international instruments or treaties to ensure that water security is achieved for all users, whether upstream or downstream (UN-Water, 2013) along with robust conflict resolution mechanisms where competing uses occur to avoid and address disputes between states (Wouters, 2010, p. 7).

On the other hand, there are voices claiming that securitization of resources pushes them up the political agenda and makes matters related to them harder to compromise upon. This is the strand of literature that views water concepts of security in water governance as harmful to negotiations and the necessary compromises between the parties involved (e.g. Leb & Wouters, 2013). Another reason why securitization of environmental or resource issues may be detrimental has to do with the concealment of the human-induced factors in certain crises (Fischhendler & Katz, 2013). For example, Dabelko (2009) criticized the link made by some between the climate change and the genocide in Darfur as such claims, in his opinion, push aside the political and economic motivations for the fighting by setting the natural factors to the forefront. Furthermore, in research around the Nile River basin, Mekonnen (2010, p. 438) argues that discourse on 'water security' was defined as 'the right of all Nile Basin States to reliable access to and use of the Nile River System for health, agriculture, livelihoods, production and environment' and seen by some as the diplomatic victory of Egypt in sustaining the status-quo in terms of water allocation. In research on transboundary water governance in Asia, Mirumachi (2013) uncovers how security discourse can be used by more powerful countries to achieve their individual goals like project compliance.

The MRC and Security Discourse

Background and Emerging Pressures

In 1995, Thailand, Laos, Cambodia, and Viet Nam signed the Agreement on the Cooperation for the Sustainable Development of the Mekong River Basin. By doing so, the four downstream riparians agreed to cooperate in all fields of sustainable development, utilization, management, and conservation of the water and related resources of the Mekong River Basin. To implement this commitment, the 1995 Agreement establishes a RBO—the MRC, tasked with implementing the objectives of the Agreement. The MRC focuses on a broad variety of issues related to water resources governance, aimed at ensuring IWRM that brings together the different interdependent challenges in the basin. It relies on a governance structure consisting of a high-level ministerial decision-making body (Council), a more technical operationalization body (Joint Committee), and the MRC Secretariat as well as national coordinating entities (National Mekong Committees). The governance arrangements of the MRC require a collective negotiation between the various subgroups, including the MRC Council, Joint Committee, the Secretariat, and the National Mekong Committees (Sajor, Huong, & Ha, 2013).

Importantly, the two upstream riparians—China and Myanmar—are not included in the MRC. But Dialogue Meetings, held annually since 1996, provide opportunities for the MRC to exchange with the two non-member basin states, and in recent years, China has agreed to share hydrologic data and information on river flows and dam operations (Figure 1).

Broadly speaking, the MRC has been regarded as relatively successful in mitigating conflicts and maintaining cooperation in the basin (Delli Priscoli, 2009, p. 29; Ha, 2011). It is seen as having a real role to play in contributing to international cooperation and notions of regional security (Jacobs, 2002, p. 363). But it has been criticized for its heavy donor and international organization involvement (Schmeier, 2013). Some see the MRC as ineffectual, sidelined by its own member-states insisting on absolute national sovereignty, and marked by mistrust and miscommunication (Lebel et al., 2005; Sajor et al., 2013). Historically, some donors have looked

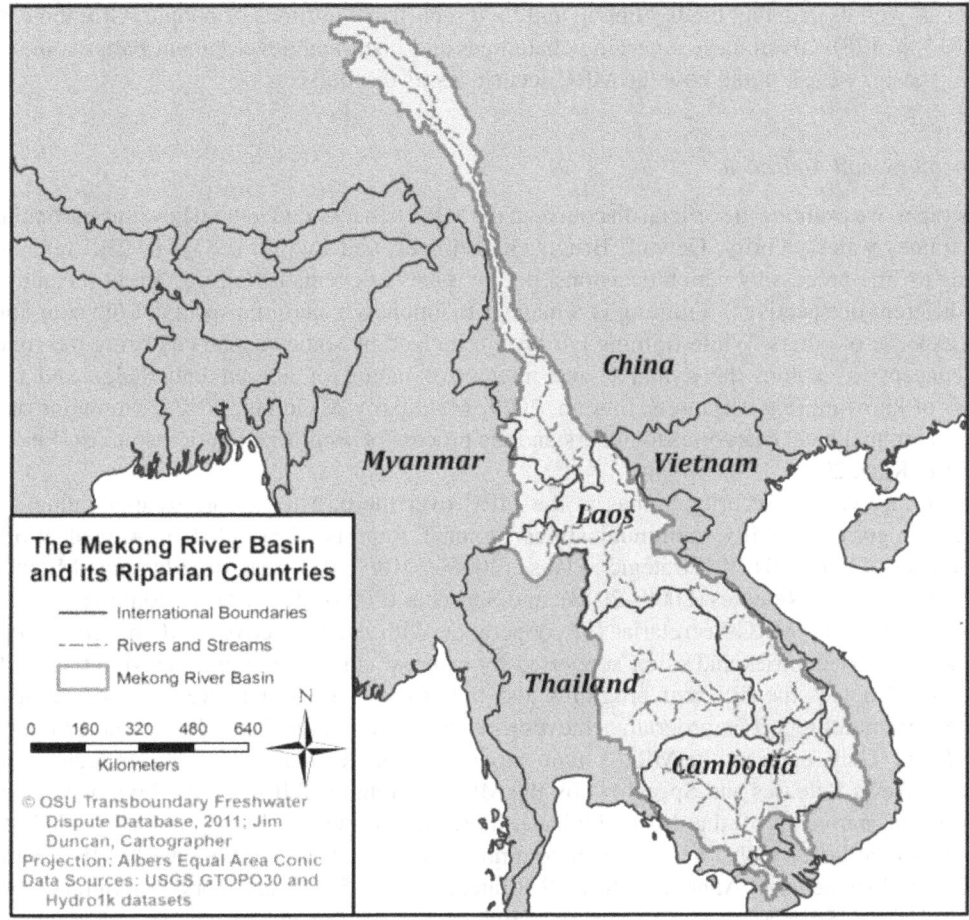

Figure 1. The Mekong River Basin.

upon the MRC more of a technical and managerial vehicle for operating projects and programs, and less as an organization for water governance in the basin (Hirsch & Jensen, 2006, p. xxii). Although stakeholder involvement has improved in recent years, the MRC is seen by some as a distant organization, inaccessible to both NGOs and communities, and reluctant to protect weaker states or elements of society vulnerable to the impacts of development (Dore & Lebel, 2010; Hirsch & Jensen, 2006, p. 51; Kirby et al., 2010). Importantly, it is not the only venue for basin development in the region; both the Greater Mekong Subregion (GMS) growth zone and the ASEAN promote economic development among the riparian states (Weatherebee, 1997).

New pressures are emerging in the Mekong River Basin, such as population growth and urbanization, and increasing demands for agricultural and fisheries production (Biswas & Seetharam, 2008a; Kirby et al., 2010). At the same time, the increasingly rapid pace of hydro-development in the upstream of the Mekong basin is threatening the integrity of the river system, posing a real concern for Lower Basin states, which are particularly dependent on the basin for livelihoods (Bakker, 1999; Cook, Fisher, Tiemann, & Vidal, 2011; Molle & Floch, 2008; Pearse-Smith, 2012; Sneddon & Fox, 2012). Although the proposed dams would provide substantial power, they are also expected to reduce biodiversity and ecosystem service values of the Lower Mekong Basin

(LMB), as well as undercut the livelihood and food security of millions of people (Grumbine & Xu, 2011, p. 178). Given these emerging challenges and historical development patterns in the Mekong basin, we ask if and how the MRC securitizes its discourse.

Methodology and Approach

In this paper, we examine the official discourse of the MRC to better understand how one particular RBO frames water security. Dewulf, Brugnach, Termeer, and Ingram (2013, p. 230) refer to framing as 'the process by which decisions, policy issues or events acquire different meanings from different perspectives'. Framing is a manner to emphasize certain aspects of the situation at the expense of others. While framing has been criticized by some scholars as being too rigid of a concept to capture the dynamic and interactive nature of human knowledge and the process of knowing (e.g. Lejano & Ingram, 2009; Mukhtarov & Gerlak, 2013), it remains one of the major analytical categories to understand the process of securitization of resources (Fischhendler & Katz, 2013).

We explore various security frames in the MRC over the past five years by examining six distinct categories of MRC communications: Annual Reports (2008–2010), Annual Work Programmes (2008–2012), Strategic Plans (2006–2010; 2011–2015); Meeting Minutes (2009–2013); News Releases (2008–2013); and Speeches (2008–2013). These communications are produced by the MRC Secretariat (in cooperation with the respective MRC programs and MRC's member countries and often supported by external consultants) and describe the work of the MRC across the different programs and initiatives. The Annual Reports and Annual Work Programmes report on program achievements year to year and signal the official activities of the MRC. They represent the MRC's main way of communicating its policies, strategies, and activities. News Releases and Speeches (by the MRC Secretariat, MRC Council members, and development partners) of the MRC indicate more select messaging of the MRC and its members to the larger global and regional communities. In total, we examined 188 MRC communication documents: 3 Annual Reports, 2 Strategic Plans, 5 Annual Work Programmes, 87 News Releases, 78 Speeches, and 13 Meeting Minutes. See the Appendix for coding data figures.

We follow an abductive logic of research, moving in an iterative manner, both inductively and deductively, between code development and from our coding data (Schwartz-Shea & Yanow, 2012). We first identified instances where 'security' language appears in the text and then coded the texts based on the categories for policy strategy and scale.[1] Our unit of analysis is the text around which the term security appears in the document. In this way, there may be many instances of security framing in one unit of analysis. There is precedence for use of the term 'security' as an indicator of issue framing in earlier research (Cook & Bakker, 2012; Detraz & Betsill, 2009; Fischhendler & Katz, 2013). In this research, we are particularly interested in how security is framed, at what scale, along with proposed policy strategies related to security.

Our Research Findings

The Framing of Security in MRC Discourse

Overall, we find many instances of security language adopted in official MRC discourse. In our analysis, we found 130 specific instances when security language was adopted by the MRC. Many diverse types of security are used including energy (six instances), livelihood (six instances), physical (six instances interpreted from port, critical services, security management systems), water (five instances), environmental (two instances), resource (three instances), human (one instance), health (one instance), and social order security (one instance). Yet, we find that the dominant security discourse is overwhelmingly food security. Food security represents 76% of the references to security, or 99 references out of 130.

The MRC formally recognized food security as a priority area in the MRC's Hua Hin Declaration of 2010, where member-states agreed to identifying and advising on the opportunities and challenges of hydropower and other infrastructure development in the Basin. Food security is also a priority under the MRC's Basin Development Planning, which prioritizes irrigated agriculture for food security, environmental and social sustainability of hydropower development, climate change adaptation strategies, and integrating basin planning into national and regulatory systems (MRC, 2011a).

According to the MRC discourse, the resources of the basin are thought to provide 'food security' for people living in the basin. Quite commonly, the Lower Mekong fisheries are tied to food security in the region. Fisheries and other aquatic plants and animals are seen as playing a 'vital role in ensuring food, income and livelihood security for many people across the Lower Mekong Basin' especially in rural areas (MRC, 2012a, p. 5). Food security is consistently linked with poverty alleviation and development in the security discourse of the MRC. Initially, we find that attention to food security was situated solely within the MRC's Fisheries Programme, but we uncover an expansion to other MRC program areas, beginning in 2009 and through 2013 (e.g. Agriculture, Irrigation and Forestry Programme; Initiative on Sustainable Hydropower, Climate Change Adaptation Initiative).

Figure 2 below displays a growing attention to security in the MRC discourse, which peaks in 2010. Food security as a frame is in decline after 2010 with other types of security emerging and filling the security discourse space, including energy, water, environmental livelihood, health, and social order security.

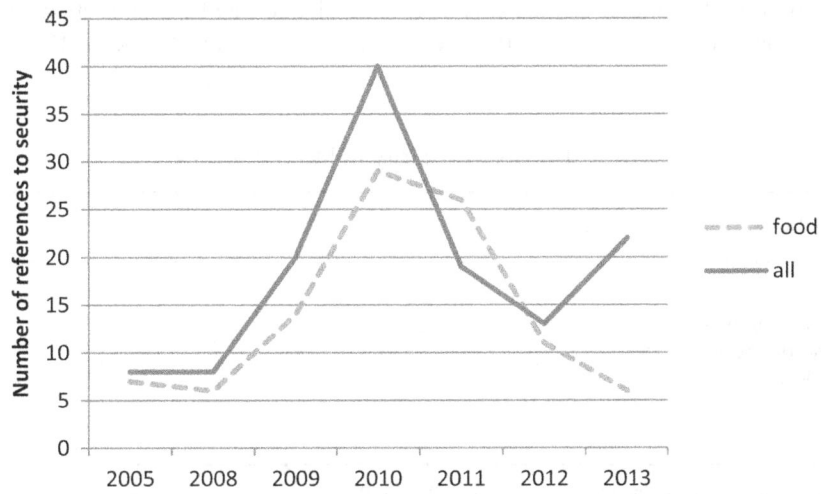

Figure 2. Security and food security over time in MRC discourse.

Perceived Threats in MRC Discourse

A variety of threats to security are identified in the MRC discourse. Climate change is invoked most as a threat to security (26 instances, or 32%). Climate change is expected to make people more vulnerable to poverty and food insecurity in the basin due to predicted changes in weather conditions such as extreme drought and floods (MRC, 2010a, p. 15). In addition to climate change language specifically, the MRC discourse also reveals use of floods and droughts as threats to security (11 instances, or 13%). Flooding in the basin puts people at risk and 'endangers' food security (MRC, 2009, p. 28). Similarly, drought will impact food security and affect the economic development of people already facing serious poverty (MRC, 2010b). When coupled with specific attention to climate change, this suggests that the greatest perceived threats to security in the MRC are climate and weather related, accounting for some 45% of the perceived threats to security. Social–economic dynamics are also seen as a threat. This includes national development, global economic crisis, and population growth. So too are dams and infrastructure projects. Table 1 reports the counts on threats to security.

Policy Strategies in MRC Discourse

In terms of proposed policy strategies, we uncover a variety of strategies related to security in the official MRC discourse. The dominant theme is heightened regional cooperation and coordination, representing some 44% of the proposed policy strategies. Addressing basin challenges will require better coordinated and collaborative responses by the basin governments, particularly in terms of implications for food security, water quality, biodiversity, and aquatic ecosystems (MRC, 2011b, p. 3). Planning based on regional cooperation will yield better results than uncoordinated and fragmented planning, according to the official MRC discourse (MRC, 2005, p. 7).

A second theme in policy strategies relates to matters of economic development, representing some 36% of the proposed policy strategies. We find that economic development takes on many dimensions. It represents more traditional resource development, including fisheries, agriculture, hydropower, and resource development. It also speaks to economic development related to economic integration, trade, and investments as well as navigation and transportation. Finally, it reinforces the need for poverty alleviation development. According to the MRC's Strategic

Table 1.   Threats to security in MRC discourse

Threats to security	Counts
Climate change	26 (32%)
Social–economic dynamics	17 (21%)
Dams and infrastructure development	16 (20%)
Floods and/or droughts	11 (13%)
Irrigation	6 (7%)
Other	6 (7%)

Notes: Counts of instances when the threat was linked with security.
Other includes lack of monitoring/information, biofuels, accidents in ports, saltwater intrusion, and fisheries decline.

Plan 2006–2010, 'Wise, environmentally sound and carefully targeted investments in the water sector hold the potential to have significant pro-poor impacts' (MRC, 2005, p. 1).

In terms of development, there is also recognition that there is need for joint determination by the LMB countries and their stakeholders at the basin level on an acceptable balance between resource development and resource protection in order to maintain ecosystem services and their contribution to food and livelihoods security (MRC, 2010c, p. 15). For example, according to the MRC, to ensure food security, it is necessary to maintain productive fisheries and enhanced aquaculture of indigenous species for increased food security and economic output (MRC, 2005, p. 7). Climate change adaptation also illustrates this third theme of policy strategies. The MRC's Climate Change and Adaptation Initiative (CCAI) is proposed as a regional initiative to address climate change adaptation challenges of the member countries to help achieve poverty alleviation and food security goals (MRC, 2009, p. 27). Climate change adaptation measures are seen as necessary to minimize poverty and food insecurity among vulnerable communities, according to the MRC (MRC, 2010d) (Table 2).

Scale and Security in MRC Discourse

Finally, we examined the use of scale in how security is framed in MRC discourse. Scale can be seen as the spatial, temporal, quantitative, or analytical dimensions used to measure, rank, or study a phenomenon (Gibson, Ostrom, & Ahn, 2000). Overall, we find most attention is given to the regional scale, representing some 48% of all scalar discourse. The problems and issues related to security are primarily regional in nature. According to the MRC, actions should be at the 'basin level' (MRC, 2010d, p. 15). Importantly, the MRC sees itself as playing an important role as a regional hub for knowledge, promoting knowledge sharing and giving the MRC a central role in in improving transboundary water resources management and water security in the region (MRC, 2010e, p. 28).

We also find significant attention to the national scale, representing some 45% of the scalar discourse. The MRC's 2011 Work Programme calls for coordinated and collaborative responses

Table 2.   Proposed policy strategies to address security

Broad strategy	Actions to meet strategy	Counts
Regional cooperation and coordination (43%)	Dialogue	43
	Knowledge sharing and capacity building	26
	IWRM	11
	Basin planning	10
Economic development (36%)	Fisheries production	17
	Agriculture production	13
	Hydropower development	12
	Economic integration, trade, and investments	9
	Socioeconomic development	8
	Poverty alleviation development	8
	Navigation and transportation	7
Resource protection (21%)	Climate change adaptation	28
	Resource protection and efficiencies	15

Note: Counts of instances when proposed policy actions were linked with security.

Table 3.   Scale and policy strategies in framing security

Scale	Counts
Global	4 (5%)
Regional	40 (48%)
National	38 (45%)
Local	2 (2%)

Note: Counts of instances when scale was linked with security.

by basin governments in terms of implications for food security, water quality, biodiversity and aquatic ecosystems to address basin challenges, notable climate change (MRC, 2011b, p. 3). We find far less attention to global and local levels, however. Local is rarely invoked to highlight a particular vulnerable community or community impacted by drought or flood conditions. Global is invoked largely to reflect collaboration with international experts around proposed policy strategies. Table 3 reports the counts on scale.

Discussion: The Many Faces of Security in MRC Discourse

In our research examining the framing of security in MRC discourse, we find food security as the dominant MRC security discourse. Food security is framed as both a goal and a growing challenge in the basin, and is consistently linked to issues of poverty reduction. Given the development status of many Mekong basin riparians, it is not altogether surprising to see linkages to development and poverty alleviation. Differences in access to critical social and political processes (Gupta & Lebel, 2010) often underline existing water insecurities, including water supply for health, and hazards such as flooding and drought (Grey & Sadoff, 2007). The focus on food security incorporates the concerns over water availability for agriculture, energy production, and fisheries. In this way, the reading of food security by MRC is more akin to the anthropocentric 'human security' discourse that dates back to the early 1990s (e.g. Homer-Dixon, 2000).

At the same time, fisheries are important in the Mekong region, especially in the delta area. With the expansion of the fisheries from the designated program on MRC Fisheries to other MRC programs, it is a manifestation of how important 'food security' is in the basin. The framing of food security frequently refers to the economic growth potential associated with development and agriculture but also with the potential for insecurity if people lack food for human consumption. The 2008 Work Programme describes food security this way:

> The Mekong River has one of the most abundant fisheries in the world. About 40 million people are engaged in the Mekong's Fishery at least part time. It has been estimated that the value of the Mekong's annual fish harvest is worth about US$ 1.4 billion at point of first sale. There is no doubt that the Mekong Fishery is important to both the livelihoods of the Basin's people—in particular the poorest segments of societies—and the broader economic growth. Fish is the most important source of animal protein for the people in the region. Together with rice, it forms the basis of the food security. (MRC, 2008, p. 4)

The MRC provides some hints as to why it focuses so heavily on food security. According to its 2012 Work Programme, (t)hroughout the basin there is evidence of a rising tide of commercial agriculture in addition to traditional, small-scale subsistence cultivation. Until the re-emergence of a 'global food crisis' in 2007, in the wake of rising commodity prices, livelihood strategies and export earnings have been the complementary ends of the agricultural policy agenda in the region. The revival of a food security agenda, and to an extent, concerns about the likely impacts of global climate change, has re-ignited interest in larger interventions in public and private irrigation and agricultural development. The recent parliamentary approval of the proposed water grid in Thailand and similar proposals for large-scale irrigation development in Cambodia are obvious examples of intended large-scale water engineering. Food security interests are likely also driving alternative 'private sector' investment in contract farming and both corporate and foreign direct investment into irrigation development, with strong interest in both Lao PDR and Cambodia. (MRC, 2012a, p. 22)

Beyond these strong links to agriculture, food security has been linked in the last few years to energy issues and sustainable development (e.g. Cherp & Jewell, 2011, p. 7). The energy security discussions have seen, for example, an increased focus on the bio-energy and especially biofuels (Asveld et al., 2011). The major tenet in the discussions on sustainable bio-energy, and partly, energy security is the availability of land in the light of the global food prices crisis in 2007–2008. In southeastern Asia, where agriculture is key, the matters of food security are especially sensitive. It is therefore not surprising that food security has been central to the discussions of the MRC. Recognition of these inextricable links between water, food, and energy in the basin across different scales and boundaries can help transform environmental and social risks into future development and security opportunities (Grumbine, Dore, & Xu, 2012, p. 96).

The emergence of the 'nexus' calls for a broad reading of water security. This doctrine claims that water resources should not be treated in isolation, as if independent of the food, climate, or energy security of individuals, communities, and states (Zeitoun, Allan, & Mohieldeen, 2010). A 'broad' approach to water security emphasizes biophysical security measured mostly in water quantity and quality, based on political, socioeconomic and ecological issues, including negotiations and human security (Bakker, 2012; Zeitoun et al., 2010). The nexus, however, is not universally accepted as a helpful concept. A number of authors advocate for the more 'narrow' definition of water and other types of security for operationalization and benchmarking of the progress (Biswas & Seetharam, 2008b; Lautze & Manthrithilake, 2012).

Overall, in the MRC context, food, water, and energy security outstrip environmental security, as might be expected in the areas where subsistence agriculture and poverty are widely spread. In the earlier years studied, between 2005 and 2009, in addition to food security, we observe reference to just a few types of security, including human, resource, and physical security. Then beginning in 2010 and through 2013, we find many new types of security framed in the MRC discourse, including energy, water, and environmental security as well as livelihood, health, and social order security. These trends are in line with global developments in the discourse around security, where the connection between water and energy security is gaining in attention (Hellegers, Zilberman, Steduto, & McCornick, 2008; Hightower, 2012). For example, the Bonn 2011 Nexus conference established a platform on *Water, Energy, and Food Security* sponsored by The Federal Government of Germany (FRG, 2013). A number of initiatives take place under this theme, including assistance to developing countries in water planning in accordance with the water–energy–food nexus tailored for their specific needs. The links between water security and energy security are further widely articulated in the recent Global Energy Assessment, mostly through the pathway of multiple uses of water in which energy generation and cultivation of biocrops play an increasing role (GEA, 2013). Given that frames are often inflexible and rigid (Lejano & Ingram, 2009), we might expect food security to continue to be dominant security frame in the region.

A variety of threats are responsible for the perceived insecurities in the basin by the MRC. Climate change and floods and droughts represent some 45% of the perceived threats. But social–economic dynamics and dams and infrastructure development also pose a threat to security. On the one hand, dams are seen as a tool to help ensure greater food security and energy security, and to improve development (MRC, 2012b). Yet, we also uncover recognition of the negative consequences of this pursuit of energy security. The LMB 20-Year Plan Scenario with the proposed LMB mainstream dams is seen as severely affecting capture fisheries production in Cambodia and Vietnam, which would negatively impact the food security and livelihoods of people in Lao People's Democratic Republic (PDR) and in Cambodia (MRC, 2010f). In a joint Development Statement in 2013, development partners stated building dams on mainstream of the Mekong may 'irrevocably change the river and hence constitute a challenge for food security, sustainable development, and biodiversity conservation' (MRC, 2013).

In terms of policy strategies, we uncover themes around regional cooperation and coordination, economic development, and resource protection. Indeed, we might expect to find multiple strategies to respond to a multifaceted problem characterized by a diverse set of threats. Such situations have sanctioned a holistic approach to planning and implementation, as promoted by the concept of IWRM. It is not altogether surprising that the threats to livelihoods and various types of security in the region have been proposed to tackle with such actions as integration and coordination. The more important question is how this is to be achieved at multiple levels of governance, such as the regional, national, and local, important questions that reach beyond the focus of our study here.

We find that much of the security discourse has been used by the MRC to underline the threats of catastrophic events as well as slow and chronic resource degradation. The security discourse in our case study is therefore consistent with the alarming rhetoric in the climate discourse (Risbey, 2008), and identified in UNCSD security discourse (Katz & Fischendler, 2012). The effectiveness of such language in promoting action toward solutions, however, is debatable (Malnes, 2008). Some research suggests, for example, that if the rights, needs, and risks of vulnerable groups are excluded from decision-making about adaptation policies, insecurities could be exacerbated by not only the original threat but also by the proposed solutions (Lebel, 2007; Paavola & Adger, 2006).

To help facilitate greater transboundary regional cooperation, the MRC places itself at the center of water governance. In terms of the scalar importance of proposed policy strategies, it is not surprising that actions are proposed at the transboundary regional scale where MRC acts as a knowledge hub. The MRC has an important role to play in terms of data information and exchange, water-use monitoring, real-time flood forecasting, maintenance of flows along the river, and water quality (Grumbine et al., 2012, p. 96). At that scale, climate change is seen as a greatest challenge to regional security. For example, the MRC emphasizes exchange visits and sharing of lessons between the basin-wide pilot projects related to wetlands, biodiversity, flood, drought, food security, and sustainable hydropower (MRC, 2012a, pp. 120–121). This echoes earlier research on the MRC that suggests that there has been a rescaling of human–environmental interactions in the basin to privilege regional boundaries and undervalue local resource use (Bakker, 1999). It also reinforces earlier research uncovering a regional frame by the MRC in terms of climate change discourse (Gerlak & Schmeier, 2014). The hydropolitics of the basin are thought to privilege regional development (Hirsch, 2001; Lebel et al., 2005; Molle, Mollinga, & Wester, 2009; Sneddon and Fox, 2006). Suhardiman, Giordano, & Molle (2011) argue there is a scalar disconnect between the regional and national decision-making landscapes in transboundary water governance in the region; the MRC and country members proceed with their conflicting development plans and reproduce the current disconnect between the national and regional-level decision-making landscape, without displaying this tension in public.

Yet, many of the perceived threats, including, population growth, development (agriculture and hydropower), and threats to fisheries are essentially national or local level issues. While the ultimate solutions to these challenges may need to be regionally coordinated, they must be targeted at national and local levels. In cases where transboundary decisions impact water supply, and water and food security, more targeted attention to and engagement with the local scale is critical (Arthur, Friend, & Dubois, 2011; Chomchai et al., 2005; Grumbine et al., 2012; Maganda & Hoff, 2014). With this narrow focus on the regional, the MRC misses out on the richness of scale that characterizes the Mekong River Basin (Lebel et al., 2005). This narrow focus on scale may ultimately hinder implementation of MRC programs, like the MRC's Climate Change Adaptation Initiative, which demands national and local implementation and buy-in (Gerlak & Schmeier, 2014). It is also necessary if the MRC is going to fulfill its decentralization and riparianization agenda, including efforts to overcome weak links between the basin level and national levels (Schmeier, 2013).

Taken together, these findings suggest the use of the term 'security' in the MRC discourse to be broadly multi-dimensional. Many types of security are embraced, and often they are discussed simultaneously (e.g. MRC, 2011c). This includes the use of a variety of types of security, including water security, food security, human security, resource security, energy security, and environmental security. In the case of water, for example, security is framed in terms of regional cooperation and climate change adaptation. In terms of food, securing reliable supplies is often associated with ensuring development and poverty alleviation, or restraining the hydropower development. This suggests a broad use of security in MRC discourse suggesting that security is not independent of food, climate, or energy and has a strong link to developmental concerns. This finding allows us to claim that the concept of security is highly malleable in the context of the MRC and has taken the broad definition and conceptualization rather than the narrow one. Yet, our research suggests that this is more nuanced than originally expected, as various conflicts within the types of security emerged. This view of security requires attention to multiple threats and potential policy strategies, as our findings suggest. The calls for an integrated and collaborative approach are an expected policy strategy in a situation when various types of 'security' may come in conflict with each other, such as energy security as linked to the expansion of the hydropower and the food security as linked to the productivity of fisheries. In this way, the rising global discourse on water security may reinvigorate the older discourse of IWRM.

The downside of such a broad approach may ultimately dilute the use of the concept of security where decision-makers remain unclear about how to prioritize among competing interests and issues. If security is everything, perhaps it is nothing. Such a universal meaning that security is taking up is nevertheless crucial in terms of structuration of social relations in the MRC forum. In Laclau's (2000, p. 58) words, such a malleable concept 'through its very emptiness, produces a series of crucial effects in the structuration/restructuration of social relations'. Ultimately, it may just be that the MRC or individual actors within the organization use security as a symbolic and rhetorical tool to justify pre-existing goals and actions (Fischhendler & Katz, 2012; Haas, 2002).

Our findings offer a number of lessons for cross-border water governance. Basin organizations or other joint bodies can help facilitate dialogue, sharing of information and conduct of assessment (Lebel, Xu, Bastakoti, & Lamba, 2010, p. 369). Anticipatory policies and actions to adapt and improve adaptive capacity to the transboundary impacts of changes in water-use, land-use, and climate on water resources and services are still in their infancy; but several problem-framing discourses are emerging that have longer term implications for water governance. It is not yet clear how these competing policy frames will evolve in Asia. Much will depend on how systems of water governance develop. Public scrutiny of how governments in Asia plan to adapt to climate change in the water sector—on how risks of not enough and too much water are dealt with—will need to continue to help sort out those projects and strategies which are driven primarily by political benefits from those which actually contribute to building adaptive capacities and maintaining social–ecological resilience.

Achieving the required institutional flexibility in current RBOs and planning processes is going to be difficult and take time (Biswas & Seetharam, 2008b). Basins will not be the sole, or even primary, unit of water management or decision-making (Molle, 2008). There are some voices that claim that transboundary water cooperation is not necessarily done through RBOs, and may even be harmed by that (Fischhendler & Feitelson, 2005). While we do not suggest that the work of MRC has been counterproductive in terms of water resources management in the region, however broadly this may be defined, we call attention to the need for a multiple-level and a multi-venue approach to identifying and resolving concerns over security.

Conclusions

In this paper, we examined how security is framed by the MRC, a well-established international RBO that promotes cross-border governance of water and related resources. We find that food security is the dominant security frame in the MRC, characterized by an anthropocentric framing of security that places emphasis on water for human needs like agriculture, energy production, and fisheries. In this way, the reading of food security by MRC is more akin to the anthropocentric 'human security' discourse that dates back to the early 1990s. But we also uncover a growing multi-dimensional approach to security in the MRC's discourse, illustrated by a variety of types of security embraced and a heightened attention to new forms of security, like energy and water security, and the linkages between these types of security. Security has many faces in the MRC, in terms of the types of security embraced but also in terms of the perceived threats to security, and the diverse policy strategies to mitigate these challenges. Security is framed largely at the regional scale with the MRC poised to facilitate cooperation, promote economic development, and protection of natural resources. Overall, we uncover an emphasis on security and the links it brings to food and energy security at the regional level of water governance.

We find that some new frames are emerging to potentially threaten the dominant frame of food security. Similarly, the national frame is more widely adopted than expected based on our reading of earlier research examining the politics of scale in basin governance. Narrow frames around scale may serve to challenge implementation of MRC programs and diminish its efforts to strengthen national and local linkages. The future research may inquire into the issues of motivation behind the politics of scale involved in the MRC discussions in order to understand to what extent the use of security jargon has to do with self-establishment of the MRC and its re-invention of self-identity vis-à-vis the national member-states and international stakeholders.

We would be remiss if we failed to acknowledge the limitations of our research. Importantly, our examination of discourse within the MRC likely does not fully capture the debate and construction of security within the basin. In part, this is related to the fact that the MRC remains politically disconnected, in part by the nature of its design and structure, from the politics of its member-states and relevant constituencies. Furthermore, our research does not consider emerging synergies and relations with other regional organizations like ASEAN and GMS. Future research may well tackle these connections and inter-relationships. It may also include interviews to gain a more qualitative interpretation from MRC officials and stakeholders in the region to the versatile use of the term 'security'. Interviews might allow us to further probe linkages between various uses of 'security' and implications of such uses on policy strategies proposed. For instance, to what extent do national and local stakeholders see security challenges, threats, and policy strategies framed as true issues of national concern or as mere implementers of regional goals? In addition, future research might well investigate and compare in other RBO cases to better understand the use and adoption of 'security' discourse in other river basin contexts to better understand the contextual nature of security discourse in cross-border river basin governance.

Disclosure Statement

No potential conflict of interest was reported by the authors.

Note

1. Our approach to intercoder reliability was to achieve consensus. To do so, two coders coded all documents and resolved any coding discrepancies or differences in coding. All coded minutes were reviewed for data entry errors and calculation errors in line counts.

References

Adeel, Z. (2012). A human development approach to water security. In H. Bigas (Ed.), *The global water crisis: Addressing an urgent security issue* (pp. 70–75). Papers for the InterAction Council, 2011–2012. Hamilton: UNU-INWEH.

Allan, J. (1995). Water in the Middle East and in Israel-Palestine: Some local and global issues. In M. Haddad and E. Feitelson (Eds.), *Joint management of shared aquifers*Molle, Mollinga, & Wester, (pp. 31–44) (pp. 31–44). Jerusalem: Palestine Consultancy Group and the Truman Research Institute of Hebrew University.

Arthur, R., Friend, R., & Dubois, M. (2011). Fisheries, nutrition and regional development pathways: Reasserting food rights. In K. Lazarus, N. Badenoch, N. Dao, & B. Resurreccion (Eds.), *Water rights and social justice in the Mekong region* (pp. 149–166). London: Earthscan.

Asveld, L., van Est, R., & Stemerding, D. (2011). *Getting to the core of the bio-economy: A perspective on the sustainable promise of biomass*. The Hague: Rathenau Institute.

Bakker K. (1999). The politics of hydropower: Developing the Mekong. *Political Geography, 18*, 209–232.

Bakker, K. (2012). Water security: Research challenges and opportunities. *Science, 337*(6097), 914–915.

Biswas, A. (1999, November). Opening remarks. Proceedings of an "International workshop on water based development projects: Experiences in the world". GAP Regional Development Administration, Sanliurfa.

Biswas, A. K., & Seetharam, K. (2008a). Asian water development outlook, 2007: Achieving water security for Asia. *Water Resources Development, 24*, 145–176.

Biswas, A., & Seetharam, K. (2008b). Achieving water security for Asia. *International Journal of Water Resources Development, 24*(1), 145–176.

Blakeney, M. (2009). *Intellectual property rights and food security*. Cambridge, MA: CABI.

Bogardi, J. J., Dudgeon, D., Lawford, R., Flinkerbusch, E., Meyn, A., Pahl-Wostl, C., ... Vorosmarty, C. (2012). Water security for a planet under pressure: Interconnected challenges of a changing world call for sustainable solutions. *Current Opinion in Environmental Sustainability, 4*, 35–43.

Brzoska, M. (2009). The securitization of climate change and the power of conceptions of security. *Security Peace, 27*(3), 137–145.

Burton, M., & Molden, D. (2005). Making sound decisions: Information needs for basin water management. In M. Svendsen (Ed.), *Irrigation and river basin management: Options for governance and institutions* (pp. 51–74). Cambridge, MA: CABI.

Buzan, B. (2004). A reductionist, idealistic notion that adds little analytical value, in special section: What is human security? *Security Dialogue, 35*(3), 369–370.

Buzan, B., & Hansen, L. (2009). *The evolution of international security studies.* Cambridge: University Press.

Buzan, B., Waevar, O., & de Wilde, J. (1998). *Security: A new framework for analysis.* London: Lynne Rienner.

Cherp, A., & Jewell, J. (2011). The three perspectives on energy security: Intellectual history, disciplinary roots and the potential for integration. *Current Opinion in Environmental Sustainability, 3,* 202–212.

Chomchai, P. (2005). Public participation in watershed management in theory and Practice: A Mekong River Basin perspective. In C. Bruch, L. Jansky, Mikiyasu, & K. A. Salewicz (Eds.), *Public participation in the governance of international freshwater resources* (pp. 139–155). New York: United Nations University Press.

Conca, K., Wu, F., & Ciqi, M. (2006). Global regime formation or complex institution building? The principled content of international river agreements. *International Studies Quarterly, 50,* 263–285.

Cook, C., & Bakker, K. (2012). Water security: Debating an emerging paradigm. *Global Environmental Change, 22*(1), 94–102.

Cook, S., Fisher, M., Tiemann, T., & Vidal, A. (2011). Water, food and poverty: Global- and basin-scale analysis. *Water International, 36*(1), 1–16.

Dabelko, G. (2009, August 24) Avoid hyperbole, oversimplification when climate and security meet. *Bulletin of the Atomic Scientists.*

Delli Priscoli, J. (2009). River basin organizations. In J. Delli Priscoli & A. T. Wolf (Eds.), *Managing and transforming water conflicts* (pp. 135–168). Cambridge: Cambridge University Press.

Detraz, N., & Betsill, M. (2009). Climate change and environmental security: For whom the discourse shifts. *International Studies Perspectives, 10,* 303–320.

Dewulf, A., Brugnach, M., Termeer, C., & Ingram, H. (2013). Bridging knowledge frames and networks in climate and water governance. In J. Edelenbos, N. Bressers, & P. Scholten (Eds.), *Water governance as connective capacity* (pp. 230–247). Surrey: Ashgate.

Dore, J., & Lebel, L. (2010). Deliberation and scale in Mekong region water governance. *Environmental Management, 46*(1), 60–80.

Falkenmark, M. 2013. Growing water scarcity in agriculture: Future challenge to global water security. *Philosophical Transactions of the Royal Society A: Mathematical, Physical & Engineering Sciences, 371,* 1–14.

Falkenmark, M., & Lundquist, J. (1998). Towards water security: Political determination and human adaptation crucial. *Natural Resources Forum, 22*(1), 37–51.

Fischhendler, I., & Katz, D. (2013). The use of "security" jargon in sustainable development discourse: Evidence from UN Commission on Sustainable Development. *International Environmental Agreements: Politics, Law and Economics, 11*(3), 321–342.

Fischhendler, I., & Feitelson, E. (2005). The formation and viability of a non-basin water management: The US–Canada case. *Geoforum, 36*(6), 792–804.

Fischhendler, I., & Katz, D. (2012). The use of "security" jargon in sustainable development discourse: Evidence from UN Commission on Sustainable Development. *International Environmental Agreements: Politics, Law and Economics, 11*(3), 321–342.

Gerlak, A. K., & Grant, K. A. (2009). The correlates of cooperative institutions for international rivers. In T. J. Volgy, Z. Šabič, P. Roter, & A. K. Gerlak (Eds.), *Mapping the new world order* (pp. 114–147). Oxford: Wiley-Blackwell Publishers.

Gerlak, A. and Mukhtarov, F. (2015). 'Ways of knowing' water: Integrated water resources management and water security as complementary discourses. *International Environmental Agreements: Politics, Law and Economics, 15*(3): 257–272.

Gerlak, A. K., & Schmeier, S. (2014). Climate change and transboundary waters: A study of discourse in the Mekong River Basin. *The Journal of Environment Development, 23,* 358–386.

Gerlak A. K., & Wilder, M. (2012). Exploring the textured landscape of water insecurity and the human right to water. *Environment: Science and Policy for Sustainable Development, 54*(2), 4–17.

Gibson, C., Ostrom, E., & Ahn, T. K. (2000). The concept of scale and the human dimensions of global change: A survey. *Ecological Economics, 32,* 217–239.

Global Energy Assessment (GEA) (2013). *Towards a sustainable future: Key findings, summary for policy makers.* Technical summary. Cambridge, New York and Laxenburg: Cambridge University Press and International Institute for Applied Systems Analysis.

Grey, D., & Sadoff, C. W. (2007). Sink or swim? Water security for growth and development. *Water Policy, 9*(6), 545–571.

Grumbine, R. E., Dore, J., & Xu, J. (2012). Mekong hydropower: Drivers of change and governance challenges. *Frontiers in Ecology and Environment, 10*(2), 91–98.

Grumbine, R. E., & Xu, J. (2011). Mekong hydropower development. *Science, 332,* 178–179.

Gupta, J., & Lebel, L. (2010). Access and allocation in earth system governance: Water and climate change compared. *International Environmental Agreements: Politics, Law and Economics, 10*(4), 377–395.

Ha, M. L. (2011). The role of regional institutions in sustainable development: A review of the Mekong river commission's first 15 years. *Consilience: The Journal of Sustainable Development, 5*(1), 125–140.

Haas, P. M. (2002). Constructing environmental conflicts from resource scarcity. *Global Environmental Politics, 2*(1), 1–11.

Harrison, J.L. (2006). 'Accidents' and invisibilities: Scaled discourse and the naturalization of regulatory neglect in California's pesticide drift conflict. *Political Geography, 25*, 506–529.

Hellegers, P., Zilberman, D., Steduto, P., & McCornick, P. (2008). Interactions between water, energy, food and environment: Evolving perspectives and policy issues. *Water Policy, 10*(S1), 1–10.

Hightower, M. (2012). Water impacts on energy security and reliability. In H. Bigas (Ed.), *The global water crisis: Addressing an urgent security issue* (pp. 18–25). Papers for the InterAction Council, 2011–2012. Hamilton: UNU-INWEH.

Hirsch, P. (2001). Globalisation, regionalisation and local voices: The Asian Development Bank and re-scaled politics of environment in the Mekong region. *Singapore Journal of Tropical Geography, 22*(3), 237–251.

Hirsch, P., & Jensen, K. M. (2006). *National interests and transboundary water management in the Mekong.* Sydney: Australian Mekong Resource Centre and Danida.

Homer-Dixon, T. (2000). *Environment, scarcity, and violence.* Princeton, NJ: Princeton University Press.

Hughes, L. (2009). The four R's of energy security. *Energy Policy, 37*(6), 2459–2461.

Jacobs, J. W. (2002). The Mekong river commission: Transboundary water resources planning and regional security. *The Geographical Journal, 168*, 354–364.

Kirby, M., Krittasudthacheewa, C., Mainuddin, M., Kemp-Benedict, E., Swartz, C., de la Rosa, E. (2010). The Mekong: A diverse basin facing the tensions of development. *Water International, 35*(5), 573–593.

Laclau, E. (2000). Identity and hegemony: The role of universality in the constitution of political logics. In J. Butler, E. Laclau, & S. Zizek (Eds.), *Contingency, hegemony, universality* (pp. 44–89). London: Verso.

Lautze, J., & Manthrithilake, H. (2012). Water security: Old concepts, new package, what value? *Natural Resources Forum, 36*(2), 76–87.

Leb, L., & Wouters, P. (2013). The water security paradox and international law: Securitization as an obstacle to achieving water security and the role of law in desecuritising the world's most precious resource. In B. Lankford, K. Bakker, M. Zeitoun, & D. Conway (Eds.), *Water security: Principles, perspectives and practices* (pp. 26–47). Abington: Routledge.

Lebel, L. (2007). Adapting to climate change. *Global Asia, 2*, 15–21.

Lebel L., Garden P., & Imamura M. (2005). Politics of scale, position and place in the governance of water resources in the Mekong region. *Ecology and Society, 10*, 18. Retrieved from http://www.ecologyandsociety.org/vol10/iss2/art18/

Lebel, L., Xu, J., Bastakoti, R. C., & Lamba, A. (2010). Pursuits of adaptiveness in the shared rivers of Monsoon Asia. *International Environmental Agreements, 10*, 355–375.

Lejano, Raul P., & Ingram, H. (2009). Collaborative networks and new ways of knowing. *Environmental Science & Policy, 12*, 653–662.

Liotta, P. H. (2002). Boomerang effect: The convergence of national and human security. *Security Dialogue, 33*(4), 473–488.

Lopez-Gunn, E., De Stefano, L., & Llamas, M. R. (2012). The role of ethics in water and food security: Balancing utilitarian and intangible values. *Water Policy, 14*, 89–105.

Maganda, C., & Hoff, H. (2014). Water security in cross-border regions: What relevance for federal human security regimes. In D. Garrick, G. R. M. Anderson, D. Connell, & Jamie. Pittock, (Eds.), *Federal rivers: Managing water in multi-layered political systems* (pp. 325–338). Cheltenham, UK: Edward Elgar.

Malnes, R. (2008). Climate science and the way we ought to think about Danger. *Environmental Politics, 17*(4), 660–672.

Mansfield, B. (2005). Beyond rescaling: Reintegrating the "national" as a dimension of scalar relations. *Progress in Human Geography, 29*, 458–473.

Marty, F. (2001). *Managing international rivers: Problems, politics, and institutions.* Bern: Peter Lang.

Mekong River Commission (MRC). (2005). *Strategic plan, 2006–2010.* Vientiane, Lao PDR: Mekong River Commission.

Mekong River Commission (MRC). (2008). *Work programme.* Vientiane, Lao PDR: Mekong River Commission.

Mekong River Commission (MRC). (2009). *Annual report.* Vientiane, Lao PDR: Mekong River Commission.

Mekong River Commission (MRC). (2010a). *Annual report.* Vientiane, Lao PDR: Mekong River Commission.

Mekong River Commission (MRC). (2010b, February 26). *Drought conditions cause low Mekong water flow.* Vientiane, Lao PDR: Mekong River Commission.

Mekong River Commission (MRC). (2010c). *Strategic plan, 2011–2015*. Vientiane, Lao PDR: Mekong River Commission.

Mekong River Commission (MRC). (2010d, June 18). *Informal donor meeting 2010*. Vientiane, Lao PDR: Mekong River Commission.

Mekong River Commission (MRC). (2010e, March 2). *Statement by Mme*. Monemany Nhoybouakong permanent secretary water resources & environment administration member of the MRC joint committee for the Lao PDR chairperson of the MRC joint committee for 2009/2010 - The 31st meeting of the MRC joint committee. Luang Prabang, Lao PDR.

Mekong River Commission (MRC). (2010f, March 2). *The 31st meeting of the MRC joint committee*. Vientiane, Lao PDR: Author.

Mekong River Commission (MRC). (2011a, October 26). *Switzerland makes new financial contribution to MRC's basin development and environment programmes*. Vientiane, Lao PDR: Mekong River Commission.

Mekong River Commission (MRC). (2011b, March 8). *Finland renews its support for sustainable hydropower development, knowledge management and capacity*. Vientiane, Lao PDR: Mekong River Commission.

Mekong River Commission (MRC). (2011c). *Work programme*. Vientiane, Lao PDR: Mekong River Commission.

Mekong River Commission (MRC). (2012a). *Work programme*. Vientiane, Lao PDR: Mekong River Commission.

Mekong River Commission (MRC). (2012b, May 3). *Conference calls for innovative solutions for water, energy and food sectors*. Vientiane, Lao PDR: Mekong River Commission.

Mekong River Commission (MRC). (2013, January 17). *Joint development partner statement*. 19th MRC Council Meeting, Vientiane, Lao PDR.

Mekonnen, D. (2010). The Nile basin cooperative framework agreement negotiations and the adoption of a 'water security' paradigm: Flight into obscurity or a logical Cul-de-sac? *The European Journal of International Law, 21*, 421–440.

Milliken, J. (1999). The study of discourse in international relations: A critique of research and methods. *European Journal of International Relations, 5*(2), 225–254.

Mirumachi, N. (2010, February 17–20). Theory vs. Policy? Connecting scholars and practitioners. River development institutions and discourses of water scarcity. International Studies Association Annual Convention, New Orleans, LA, USA.

Mirumachi, N. (2013). Securitising shared waters: An analysis of the hydropolitical context of the Tanakpur Barrage project between Nepal and India. *The Geographical Journal, 179*(4), 309–319.

Molle, F. (2008). Nirvana concepts, narratives and policy models: Insights from the water sector. *Water Alternatives, 1*, 23–40.

Molle F., & Floch P. (2008). Megaprojects and social and environmental changes: The case of the Thai water grid. *Ambio, 37*, 199–204.

Molle, F., Mollinga, P.P., & Wester, P. (2009). Hydraulic bureaucracies and the hydraulic mission: Flows of water, flows of power. *Water Alternatives 2*(3), 328–349.

Molle, F., & Wester, P. (Eds.). (2009). *River basin trajectories: Societies, environments and development*. Cambridge: CAB International.

Mukhtarov, F. (2007, September). *Global water governance and the concept of legitimacy*. Proceedings of the GRSC/GARNET International Conference on "Pathways to Legitimacy", University of Warwick.

Mukhtarov, F. (2009). *The hegemony of integrated water resources management: A study of policy translation in England, Turkey and Kazakhstan* (doctoral thesis). Department of Environmental Sciences and Policy, Central European University, Budapest.

Mukhtarov, F. (2013) Translating water policy innovations into Kazakhstan: The importance of context. In C. De Boer, J. Vinke-de Kruijf, G. Özerol, and H. Bressers (Eds.), *Water governance, policy and knowledge transfer: international studies on contextual water management* (pp. 113–128). London, UK: Earthscan.

Mukhtarov, F., & Cherp, A. (2014). *Water security. Oxford bibliographies*. Oxford, UK: Oxford University Press.

Mukhtarov, F., & Gerlak, A. K. (2013). River basin organizations in the global water discourse: An exploration of agency and strategy. *Global Governance, 1*, 307–326.

Paavola, J., & Adger, N. W. (2006). Fair adaptation to climate change. *Ecological Economics, 56*, 594–609.

Paris, R. (2001). Human security: Paradigm shift or hot air? *International Security, 26*(2), 87–102.

Pearse-Smith, S. W. D. (2012). 'Water war' in the Mekong Basin? *Asia Pacific Viewpoint, 53*, 147–162.

Risbey, J. S. (2008). The new climate discourse: Alarmist or alarming? *Global Environmental Change, 18*(1), 26–37.

Sajor, E. E., Huong, L. T. T., & Ha, N. P. N. (2013). *Challenges in developing a basin-wide management approach in the Lower Mekong*. Mekong Project 4 on water governance, challenge program for water and food Mekong. Pathumthani, Thailand: Asian Institute of Technology.

Schmeier, S. (2013). *Governing international watercourses. River basin organizations and the sustainable governance of internationally shared rivers and lakes*. London: Routledge.

Schmeier, S., Gerlak, A. K., & Blumstein, S. (2015). Clearing the muddy waters of shared watercourses governance: conceptualizing international River Basin Organizations. *International Environmental Agreements*. doi: 10.1007/s10784-015-9287-4.

Schwartz-Shea, P., & Yanow, D. (2012). *Interpretive research design: Concepts and processes*. London: Routledge.

Sneddon, C., & Fox, C. (2006). Rethinking transboundary waters: A critical hydropolitics of the Mekong basin. *Political Geography*, *25*, 181–202.

Sneddon, C., & Fox, C. (2012). Inland capture fisheries and large river systems: A political economy of Mekong fisheries. *Journal of Agrarian Change*, *12*, 279–299.

Suhardiman D., Giordano M., & Molle F. (2011). Scalar disconnect: The logic of transboundary water governance in the mekong, society & natural resources. *An International Journal 25*(6): 1–15.

Tarlock, A., & Wouters, P. (2009). Reframing the water security dialogue. *Journal of Water Law*, *20*(1), 53–60.

The Federal Republic of Germany (FRG). (2013) *The water, energy and food security resource platform: Nexus*, online. Retrieved September 18, 2014, from http://www.water-energy-food.org/

UN-Water. (2009). *Climate change adaptation. The pivotal role of water*. Policy Brief. Retrieved September 18, 2014, from http://www.unwater.org/downloads/unw_ccpol_web.pdf

UN-Water. (2013). *Water security & the global water agenda: A UN-water analytical brief*. Ontario: United Nations University.

Van Harten, M. (2002). Europe's troubled waters. A role for the OSCE: The case of the Kura-Araks. *Helsinki Monitor*, *13*(4), 338–349.

Warner, J. (2006). More sustainable participation? Multi-stakeholder platforms for integrated catchment management. *Journal of Water Resources Development*, *22* (1), 15–35.

Weatherebee, D. E. (1997). Cooperation and conflict in the Mekong River Basin. *Studies in Conflict and Terrorism*, *20*, 167–184.

Williams, M. (2007). *Culture and security: Symbolic power and the politics of international security*. London: Routledge.

Wolf, A.T. (2007). Shared waters: Conflict and cooperation. *Annual Review of Environment and Resources*, *32*, 31–66.

Wouters, P. (2010). *Water security: Global, regional and local challenges*. London, UK: Institute for Public Policy Research.

Wouters, P., Vinogradov, S., & Magis, B. O. (2009). Water security, hydrosolidarity, and international law: A river runs through it. *Yearbook of International Environmental Law*, *19*, 97–134.

Zeitoun, M., Allan, J. A., & Mohieldeen, Y. (2010). Virtual water 'flows' of the Nile Basin, 1998–2004: A first approximation and implications for water security. *Global Environmental Change*, *20*, 229–242.

Zeitoun, M., Eid Sabbagh, K., Talhami, M., & Dajani, M. (2013). Hydro-hegemony in the upper Jordan waterscape: Control and use of the flows. *Water Alternatives*, *6*, 86–106.

Zeitoun, M., & Warner, J. (2006). Hydro-hegemony: A framework for analysis of trans-boundary water conflicts. *Water Policy*, *8*(5), 435–460.

Appendix. Coding Data

Document	Number of unique instances of attention to security	Number of different frames invoked
Annual Work Programme, 2012	31	13
Annual Work Programme, 2011	38	16
Annual Work Programme, 2010	26	14
Annual Work Programme, 2009	8	4
Annual Work Programme, 2008	7	5
Subtotal	110	52

(Continued)

Appendix. Continued

Document	Number of unique instances of attention to security	Number of different frames invoked
Annual Report, 2010	16	12
Annual Report, 2009	17	11
Annual Report, 2008	12	8
Subtotal	45	31
Strategic Plan (2006–2010), 2005	31	14
Strategic Plan (2011–2015), 2011	59	23
Subtotal	90	37
Total	245	92

Document	Total no. of documents	No. of unique documents with attention to climate change (and % of total documents)	Number of unique instances of attention to security	Number of different frames invoked
News Releases, 2013	4	1 (25%)	4	3
News Releases, 2012	6	3 (50%)	9	7
News Releases, 2011	12	1 (8%)	8	8
News Releases, 2010	26	7 (27%)	46	21
News Releases, 2009	19	3 (16%)	25	15
News Releases, 2008	20	0 (0%)	0	0
Subtotal	87	15 (17%)	92	54
Speeches, 2013	13	4 (31%)	11	8
Speeches, 2012	3	0 (0%)	0	0
Speeches, 2011	22	2 (9%)	13	10
Speeches, 2010	12	3 (25%)	25	14
Speeches, 2009	11	2 (18%)	7	7
Speeches, 2008	17	2 (12%)	7	6
Subtotal	78	13 (17%)	63	45
Meeting Minutes, 2013	1	1 (100%)	7	6
Meeting Minutes, 2012	0	0 (0%)	0	0
Meeting Minutes, 2011	6	3 (50%)	23	12
Meeting Minutes, 2010	4	4 (100%)	47	15
Meeting Minutes, 2009	2	1 (50%)	26	10
Subtotal	13	9 (69%)	103	43
Total	178	3 (21%)	258	142

Of River Linkage and Issue Linkage: Transboundary Conflict and Cooperation on the River Meuse

JEROEN FRANK WARNER

ABSTRACT *It is a truism in mainstream International Relations that issue linkage promotes regime formation and integration. The present article applies this idea to the transboundary lower river Meuse and finds its history of integration to be a tortuous one. Contextual political factors have at times promoted integration, at times fragmentation. The path towards regional integration, then, has not been not linear, but has consisted of conflict and cooperation, of (Meuse–Scheldt) river linkage and issue linkage, but also counterlinkage and non-linkage. Clearly linkage is not necessarily positive. I will argue that this does not need to be problematic, but suggest accepting more complexity in the analysis of river integration. I propose a way to create some order in the many available concepts of linkage to map out the role of linkage in integration.*

1. Introduction

Recognition of the environmental interdependence and connectivity of rivers suggests their integrated management. But how about infrastructural and political linkage? Moreover, much literature (e.g. Lindemann, 2008) assumes overall integration facilitates water integration. Integration theory holds that the 'state of bilateral relations' between riparian states, that is, the level of regional integration is important when explaining international water regime formation providing an enabling environment facilitating the development of 'issue-linkage' strategies. In security literature, a linear evolution from anarchy via regime formation to integration is presumed (e.g. Busuttil et al., 1994). The road leading there is disputed, though: in the neo-realist tradition, the existence of a strong hegemonic power is assumed to promote regime formation, in the neo-liberal institutionalist (pluralist) perspective, functionalist cost–benefit calculations and linkages are assumed to do the trick (Section 2).[1]

The validity of this linear assumption is tested in the presented contribution by delving into the shared Meuse (Maas) basin, a fresh-water body shared by the Netherlands, Belgium, and France. While France outsizes the Netherlands and Belgium, it has rarely shown much interest in the transboundary Meuse and Scheldt, so that issues of contention in practice are between Belgians and Dutch. There have long been controversies over the distribution of Meuse water, especially after construction of new navigation canals diverting water from the Meuse. Physical river linkage and political linkage in negotiations have played an important role. Documentary analysis and interviews held in 2012 with (Dutch) water managers informs this contribution.

The case study, presented in Sections 3 and analysed in Section 4, suggests that linkages and hegemonic power are not necessarily conductive to cooperation. I will therefore propose and apply a provisional, more finely grained typology of linkages (Table 2).

2. Cooperation and Integration: Issue-Areas, Linkages, and Regime Change

While International Relations tend to focus on conflict, theories of regional cooperation, and integration have proliferated after the Second World War. Mitrany (1948) suggested river basin organisations can be excellent vehicles of regional integration, a sentiment still echoed today (Jagerskog, 2013).

When predictions of inevitable regional integration based on spillover between issue-areas have proved less than successful, regime theorists then proposed to 'slice up' elements of neofunctional, separating issue-specific institutions from overall regional integration (Stokke, 1997). Transboundary water management as drivers for cooperation however remained an attractive focus, with the river Rhine Action Plan presented as a towering example.

What Schulz (1995) has called *a hydropolitical security complex* points at the interdependence of water problems (scarcity, transboundary flooding, and pollution) within a transboundary river basin which can only be resolved in the context of cooperation within that same river basin—even if the responsible managers may deem it more expedient to deny the existence of such a complex (Turton, 2002). Within such security complexes, explicit or more implicit (Euphrates–Tigris) regimes of cooperation may emerge.

Regime theory sets out to explain 'patterned behaviour' (Puchala & Hopkins, 1983) in the international domain. Regime theory starts from the assumption that politics 'can be broken down into different issues, which are dealt with by different actors and in which different power relations prevail' (Gupta, van der Wurff, & Junne, 1995). The more limited and well-defined the issue, the better chances of successful regime formation.

Regimes always deal with issue-areas, rather than single issues. Issues are defined as agenda items (note that this excludes non-issues and non-agendas, cf. Lukes, 1974/2005)). Issue density denotes the number and importance of issues within a policy space. The result is non-decomposability: separately conceived issues are in fact closely linked—you cannot affect one without affecting others. An issue area is a recognised cluster of concerns involving interdependence of both parties and issues.

In the domain of water, literature started to emerge at the turn of the century predicting an inevitable move from conflict to cooperation in transboundary water management, after a lengthy obsession with water wars. Allan's river basin trajectory, from sectoral to integrated water management, was presented as an inevitability (Turton & Meissner, 2003). River basin closure, but also flood events, may bring reflexive modernisation (see below). Linkages are expected to promote integrated river basin management (Heikkala, Schlager, & Davis, 2011). It is clear however that these 'phases' are not cast in stone (Wester & Molle, 2009) and are also subject to intersubjective construction.

Likewise, a move from anarchy to integration, from securitised conflict to cooperation in a state of 'asecurity' where violence would have become inconceivable (Oelsner, 2005) does not seem inevitable. While Buzan and Waever (2003), after Wendt (1999), identify 'Hobbesian' (enmity), 'Lockeian' (rivalry), and 'Kantian' (friendship) relations between states in security complexes, they do not predict a priori an inexorable progress from conflict to cooperation, enmity to friendship.

The three stages in each of these trajectories map quite well onto three out of Lindemann's (2008) four roads to integration: power-driven, interest-driven, and context-driven (the fourth one, knowledge-driven regime formation, falls outside the scope of this paper).

Table 1 maps out the three forms of relations between states in transboundary water relations. After that each of the attending schools of thought is briefly explained.

(A) Power-based neo-realism: conflict is normal

In realist IR, states are assumed to be the principal actors, unitary (speaking with one voices) and acting rationally: they strive for the best possible solution in relation to their capabilities—apart from military force, it may include a nation's GNP, demographic resources and natural resources (including water).

A state's key interest is national security. States act to maximise the national interest, which they claim to represent. Machiavelli even wrote that the importance of state security may justify certain act by the Prince (the State) that would be forbidden to others. The ends justify any means (separation of politics and ethics).

Interdependence is not seen as positive in Realism. Realist leaders try to reduce vulnerability; it increases dependence. Lowi (1993) suggests the odds for cooperation on transboundary water are low, especially for upstream states who have little to gain from cooperation. In the present

Table 1. Three schools of thought on transboundary water relations

Relations in Buzan's security complex	IR School (cultures of anarchy)	Relations	Likelihood of war	Key ambition	Water development stage
A. Anarchy	Neo-realism *power decides*	*Hobbesian* enmity. Dog-eat-dog fights to the death	Preparation for war	Eliminate other	Hydraulic mission
B. Mature anarchy	Neo-institution-alism *interests decide*	*Lockeian* rivalry. Generally peaceful competition; some exchange and regime formation	War cannot be ruled out, but interdependence is growing	Co-exist with Other	Reflexive modernity, move towards demand management
C. Security community	Integration-ism *context decides*	*Kantian* friendship. Integration, solidarity	War has become unthinkable	Integrate with other	Integrated water management

Source: Based on Bussutil et al. (1994), Wendt (1999), Buzan and Waever (2003), Meissner and Turton (2003).

Table 2. Types of linkages, organised in sets of attributes (by the author)

Material	Discursive	(1) Fahey (2014)
Passive	Active	(2)
Strategic	Tactical	(3) Sung (2009)[23]
Hard (coercive)	Soft (consensual	(4) Bow (2009)
Constructive	Destructive	(5a) Bow (2009)[24]
Complicating integration	Facilitating integration	(5b)
Tight	Loose	(6) Perrow (2004)
(Too) many	(Too) few	(7) Warner et al. (2010)
Horizontal	Vertical	(8) Young (2006)
Durable	Nondurable	(9)
Well-timed	Ill-timed	(10) Gupta et al. (1995)

case however the upstreamer (France) is relatively uninterested while the mid-streamer (Belgium) and downstreamer (Netherlands) are relatively interdependent.

States can coerce (threaten) or bribe (persuade with side payments) their counterparts, but given the zero-sum outcome of their preferences, it makes little sense contracting to a common institution to realise plus-sum outcomes, 'nor can states engage in mutually beneficial political exchange through issue linkage' (Legro & Moravcsik, 1999). Realists see the lack of attention to power asymmetry and hegemony as a general shortcoming in liberal regime theory (Warner, 2011; Warner & Zawahri, 2012). These asymmetries will produce unequal benefits, which always leaves the option of Best Alternative to a Negotiated Agreement (BATNA). Power disparity is only useful and positive in that it creates stability of expectations. But even when apparently cooperating, an actor may deliberately fragment the cooperative regime by separating domains, an effective way of thwarting integration (Buntaine, 2007).

In classical Realist thought, environmental issues such as water rank as low politics: economic and environmental problems, as opposed to military security and diplomatic relations. Yet the Copenhagen school of security studies, which owes a lot to Realism but also accepts constructivist maxims, have accepted that water can be high politics. Placing issues in the security domain impedes relinquishing sovereignty and promoting policy integration. 'Securitising' an issue (Buzan, Waever & de Wilde, 1998) promotes it to 'pre-political immediacy'. While water traditionally is a low politics issue, there are multiple examples of water being securitised. Security from flood and drought is key here. Floods in the Netherlands command such pre-political urgency. Saying dikes is saying security (Buzan et al., 1998). Moreover, as we shall see, economic securitisation—the vital importance of Antwerp versus Rotterdam to national economic survival, that is, ports as *pars pro toto*—has played a vital role throughout the ages in justifying uncooperative behaviour, and even today rears its head. We will approach the power-based approach from this neo-realist angle.

(B) Interest-based liberal institutionalism (Pluralism)

In pluralist thinking, the state is not unitary at all: bureaucratic politics and personalities do have a bearing. The state is a multiplicity of actors, while non-state actors may be prime movers in the international domain. There may be competition between government departments, each having their own constituency, who may forge alliance with their departmental counterparts across borders. They are the quintessential 'boundary spanners'.[2]

For neo-institutionalists, power asymmetries or hegemony are secondary; rational calculation of common interest dominates this approach. The assumption is an underlying harmony of interests among individuals. Neo-institutionalists stress the importance of perceptions, from the belief that differences can ultimately be aligned. Interdependence, then, is a good thing, since it reduces nationalism and promotes integration, and stable peace. Socioeconomics is as important as military matters: commercial, democratic and regulatory liberalism will abolish war.

Regime theory regards multiple links as a potentially *positive* quality, as it is liable to increase actor compliance in one area for fear of retaliation in other areas, and facilitates package deals.

Transboundary co-operation in Europe is indeed widespread, but as Kistin (2010) has flagged up, the depth of cooperation tends to disappoint once a treaty has been signed—there may not be much beyond be a treaty and technical exchange.

(C) Integrationalism: Context matters

The great functionalist thinker on integration, Mitrany (1965), believed in the separation of technical areas from politics and integration in other areas (spillover). The existence of multiple channels decreases the utility of military solutions. Managing interdependent relations may involve the construction of sets of rules and institutions to govern relations among actors. Differences may persist, but if an equifinal meaning can be found, alignment is possible. As a consequence, government will become smaller. International negotiation is a two-level game (Putnam, 1988): domestic politics influence international politics and vice versa. The boundary separating high and low politics is thus blurred.

This approach assumes that integration in non-water domains such as the formation of the Benelux and the European Union will spill over into cooperation over water. However, *other contextual developments* may also be at play (Lindemann, 2008), whether favourable or nonfavourable for water cooperation or non-cooperation: events such as war and natural disaster, and political developments such as federalisation can play a part. Given its decades-long process of integration, we would not expect escalation over water.

2.1. Linkages

A key driver of cooperation and integration is thought to be the creation of linkages. For Levitsky and Way (2010), the concept of linkage denotes the density of (all kinds of) ties and (all kinds of) cross-border flows. The most powerful source of linkage for them is geographic proximity. The capacity of linkage moreover has recently gained much traction in the water domain as the basis for benefit-sharing between river Riparians as a consequence of the so-called water-food-energy nexus.

International regime theory starts from the notion of complex interdependence (Keohane & Nye, 1977), characterised by absence of force, no hierarchy between issues and ultipelç levels of contact between societies. Interdependence however is asymmetrical: A depends more on B than B on A. This gives B leverage, defined as vulnerability to external pressure.[3] The anticipation of coercive linkages imposed by the stronger party may induce self-restraint on behalf of the weaker party (Bow, 2009). Coercive linkage, in this line of analysis, is tit-for-tat retaliation (directly, unambiguously) while soft linkage may be 'malign passivity' such as holding grudges (Bow, 2009).

The 'complexity' of interdependence is not only reflected in actor relations and preference structures, but also in their linkages. Sub-issues are not isolated, but tend to be structurally linked to a host of other issues. The question is however why a political actor would make active links between such sub-issues. After all, the chances of successful regime formation are believed to be higher the more limited and well-defined the issue at stake. Countries with good relations tend to discuss their issues in isolation (Le Marquand, 1977); linkage is expected in conflictive constellations. High issue density however promotes economies of scale (Keohane, 1987). Also, creating linkages means that violation of a regime will have undesirable consequences in other issue-areas (Hasenclever, Mayer, & Rittberger, 1996). Linkages, then, are efforts to break an impasse or otherwise improve one's bargaining position on a particular issue by tying it to another, unrelated issue' (Bow, 2009). 'Issue trading' can expand the range of mutually acceptable settlements. Linkages can be cooperative or coercive, and they can be prospective (promises and threats) or retrospective (rewards and retaliation). Slightly differently conceived, Bernauer (2002) suggests upstream–downstream problems are resolved by a mix of coercion, reciprocity, and linkage (exchange) politics. Where multiple types of cross-scale linkages exist concurrently, they constitute a 'polycentric system of governance'. Focusing on one such linkage only may distort the picture of the benefits of such linkages (Heikkala, Schlager, & Davis, 2011).

From a functionalist perspective, regimes reduce costs by linking issues, encouraging nesting of regimes within one another. Conscious linkage therefore is a way of exploiting interconnectedness.[4] Linkage brings cost incentives to others by increasing their costs or benefits of (non)-cooperation. But not everyone can change everyone else's costs with impunity. The credibility of a (hegemon's) threat depends on a linkee's perception of the linker's ability to withhold collaboration if linkage is refused. As Gupta et al. (1995) explain, linkages encourage compliance, for fear of retaliation in other areas, and facilitate package deals and compensation.

Regime theory assumes that power configurations change from issue to issue. But linkage itself may change these relations. The strategic inclusion of specific interests in a regime (actor linkage), even if they themselves are relatively disinterested, will influence the nature, credibility, and effectiveness of regimes.

Alternative (passive) linkages need not be the result of strategic choice, but may come about in reaction to, for example, shocks and disasters. Major changes in one regime may lead to pressures for change in other regimes. The transversal connexion of different functions and levels will easily spread to contiguous areas, increasing the potential for independent existence (linkage). Regimes and cooperation in one issue-area may arise as an unintended consequence of cooperation in another area (Haggard/Simmons, 1987).

From a functionalist perspective, the decisive aspect is whether the linkage leads to a deal. If a linkage is perceived as too coercive, that linkage can be destructive. A perceived 'negative linkage' (Bow, 2009) only complicates a bargain by unnecessarily adding an unwelcome condition.

There is a host of linkage concepts available in the literature, but not well-structured. Gupta et al. (1995), for example, identify a sequence of five types of linkage:

- Material linkages: issues that are so logically interrelated that they cannot be separated.
- Political linkages: state preoccupations: national security, regional aspirations, territorial claims.
- Bargaining linkages: artificial constructions linking unrelated issues.
- Organisational and procedural linkages: the specific place (forum, setting) or timing of negotiation may promote or hinder progress in other areas.
- Conceptual linkages: definitions convened in one context may promote progress in another area.

Young (1999) teases out four types of linkages in institutional interplay

- Functional linkages: substantive connection between activities of A and B

- Political linkages: active linkage established to link or integrate A and B
- Horizontal linkages: links and relations operating at the same level, for example, State to state
- Vertical linkages: links and relations cutting across scales, for example, state to region.

I propose to organise these linkages differently (Table 2), as contrasting sets of attributes. As part of this, a key juxtaposition I would like to emphasise in this typology is whether linkages between actors or issues are simply (but irrevocably) there, such as geographical proximity, or if they need to be deliberatively, artificially connected through an intervention to establish linkage that previously was not there? River linkage—that is, physical (water) linkage such as channelisation—is a technical, material intervention, while bargaining linkage, actor linkage, and issue linkage, which abound in regime literature, are strategic, discursive linkages for advantage in negotiation.

I will also draw particular attention to potential downsides of linkages, which showed up in studying the Meuse case, we may identify some (rare) caveats in the linkage literature. From the functionalist perspective of integration, not all linkages are 'positive', that is, leading to cooperative outcomes. A combination of issues that fail to create a positive-sum game may undermine the effectiveness of a linkage strategy (Davis, 2004). Rather than complementary and productive, links may be conflictual and counter-productive (Berkes, 2002). Indeed, Brouwer (2013, p. 157) notes that linkage will only bring social and political support for policy proposals provided if linkages with controversial issues are avoided. Also they should be empathetic 'to actually build (minimum required) coalitions to accomplish tasks, policy entrepreneurs often have to be prepared to adjust their ideas to the interests and expectations of others' (Brouwer, 2013). When a certain decision creates losers, the latter may be compensated by linking issues.

Warner, Lulofs, and Bressers (2010) and Warner (2012) identified setbacks to extensive linking. This notably concerns Christmas tree-type interlinkages—a surfeit of connections through multiple linking may make negotiations overly complex and aggravate antagonistic relations between parties. This also relates to what may be termed the issue of cohesion or intensity: are the linkages between issues or actors (made) tighter or looser? Regime theory is not too clear about whether and how these linkages can be graded, but it is clear that not all of these links are equally strong. I propose to allow for loose vs. close coupling, analogous to the concept of close and loose coupling of systems (Perrow, 1984 [1999]). The lack of alternative routes, which can lead to 'normal (diplomatic) accidents' (Perrow, 1984 [1999]), could in my view improve regime theory's explanatory power in this respect. Closely coupled systems are vulnerable systems. Environment-driven disasters such as flood and drought affecting non-resilient systems can be considered integral threats to human security. Despite European integration, water therefore continues to be seen as a life-or-death issue. Perrow (1984 [1999]) has shown that closely coupled systems with little 'slack' (seemingly redundant alternative linkages) are accidents waiting to happen—vulnerable to 'acts of God'.

We will now look at a concrete case, the transboundary river Meuse (Maas) and identify different linkages.

3. Meuse Case Study

3.1. *Introduction*

The river Meuse originates near Nancy, Northern France, and carves out a deep valley in France and Belgium before entering the Netherlands at Eijsden, over 800 km from its source. In Belgium, the Meuse encompasses most of the Belgian Ardennes and Sambre region; rainfall in the Ardennes is 1000–1200 mm, compared to only 800 mm in the Netherlands. South of Charleville-Meizieres, this central Meuse is attenuated by a lateral channel and a wide floodplain. Here however the tributaries with the steepest gradients also originate. The Meuse has 30 tributaries, some of which cross sovereign borders, so that its catchment in fact includes Luxembourg and a minor part of the German *Bundesland* of Nordrhein-Westfalen, Germany. Confluence with the Niers, which flows in from Germany, may speed up flood waves on the Dutch Meuse (de Wit et al., 2007).

As a fast-flowing river, the Meuse has threatened human lives and assets in France, Belgium, and the Netherlands. Chances of winter floods appear to be increasing (see also Tol, 2000). The Meuse is a natural border between the Netherlands and Belgium for 50 km. By Dutch standards, this Border Meuse has a steep drop, 45 cm/km, as it enters the Netherlands, after which the gradient flattens sharply, and the river loses its momentum. In the past millennium the Meuse has changed course during large floods; human technology has largely tamed the Meuse, but not this stretch of the Meuse, originally a braided river, a system of gullies with natural gradients and low-lying islands which were frequently flooded. This is an unpredictable river stretch: the mean discharge, 230 m³/s, is not a very informative figure. As the rain-fed Meuse is prone to extremes, from flash floods of 3100 m³/s (in 1993) to almost zero (25 m³/s), an effective flood warning system is no luxury (Duivenvoorden, 1997). The steep drop makes the Maas the only Dutch white water river, poorly suited to shipping; ships can use a side channel running in parallel to the Maas, the Julianakanaal. An unregulated middle stretch of a river is unusual for Europe, but pays its environmental dividends, as the absence of navigation benefits the survival of rare fish species (Figure 1).

The lower Meuse flows from through the central Netherlands before emptying into the North Sea. The Meuse is an important fresh-water resource for the canal system of the province of Noord-Brabant, providing millions of Dutchmen with drinking water and supports agricultural activities. It also enables shipping freight to various industrial towns. The Meuse is used, moreover, for hydropower on both the Belgian and Dutch (Limburg) side. As we shall see, this has a considerable impact on water levels.

Given the area's low population (Dutch Border Meuse: 15,000 inhabitants in 1998), extensive agriculture, and relatively high altitude respective to sea level, there has not been an obvious need to dike up the river Maas in Limburg. The Border Meuse however was fixed between 1860 and 1890 as a narrow, 60 m wide trench. As a result, the river speeded up and eroded the gravel river bed.

The otherwise scenic Maas has been exploited for its valuable gravel deposits, as a key economic resource. Quarrying deepened the river, creating thousands of unsightly gravel pits in Southern Limburg, filled up with water and used for pleasure boating (*Maasplassen*). Gravel digging also generates noise, dust pollution and heavy transport. Cracks in houses still evidence damage from the vibrations that come with digging. Together with nature organisations, citizens from the affected towns staged protests, bringing promises that mining would be phased out and 'greened', establishing a new linkage between industry and the environment (Warner, 2012). This plan was speeded up after a flood hit Belgium and the Netherlands in 1995 (Rosenthal, Hart, & Bezuijen, 1998). The following section sketches the wider history of Belgian–Dutch river relations on the Meuse.

3.2. *Transboundary Meuse Interaction: Ancient History and the Shadow of the Past*

The problematic relationship between the Netherlands and Belgium dates back from at least the 80-year war of independence waged with Spain (1568–1648). Both were part of the Spanish

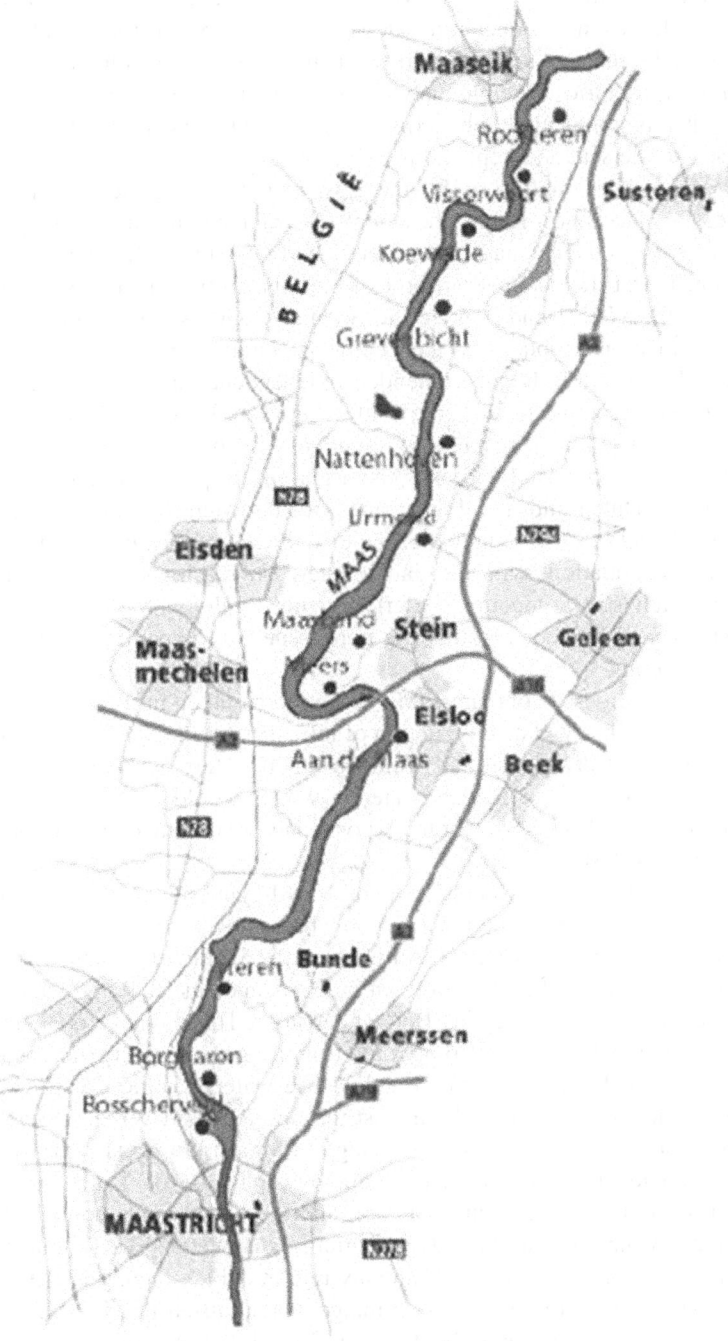

Figure 1. The Common Meuse area (www.stowa.nl).

(Hapsburg) Empire, but the Protestant North (including Flanders) rebelled against the heavy enforcement of Catholicism and declared independence in 1581, to form the United Provinces. The Spanish Kings did manage to crush Lutheranism in Antwerp, reconquer Flanders, and keep control of the lowest countries. The Flemish Low Countries had been economically superior, but suffered badly when first the Dutch rebels (in 1572), then the Spanish (in 1585) closed their port Antwerp's access to the sea, and kept it that way.

After the Peace of Westphalia (1648), the Low Countries regained independence but the blockade stayed in place until 1814. The country's fragmented water management environment had been unified under a national water department by under the French occupation, which was retained after the occupation forces were ousted. In 1815 the Southern Netherlands were assigned to the Netherlands under the Vienna Congress as a buffer against France, including the province of Limburg and the Zeeuws-Vlaanderen area, which borders the entryway to Antwerp. When the Belgians demanded independence in 1830, Limburg and Brabant supported the rebels, leading to a fierce battle between Dutch and Belgians. The Dutch responded with military force, reclaiming much of the two provinces. In 1839 Belgium gained formal independence and in 1843 the right of trespass was agreed. Not only was Limburg split into a Dutch and a Belgian part, with the Meuse as boundary; the separation treaty also crucially left to the Dutch the disputed Zeeuws-Vlaanderen territory. Whenever modern-day Dutch negotiations with Belgian counterparts reach an impasse, traumatic experiences with the Dutch still come up. Underlying this rivalry and subsequent issue linkage between the Meuse and Scheldt remains the core issue of competition between the ports of Antwerp and Rotterdam.

The term 'linkage politics' (Lohmann, 1997) came to be taken in its most literal sense in the nineteenth century, which saw the construction of infrastructure to develop shipping connection between the Rivers Scheldt, Meuse and Rhine. Until the start of the nineteenth century the Meuse had been a braided river. Excess water in winter would go into branches. Dikes and bridle paths were built for horses pulling track boats. Dikes and gravel extraction increased velocity of flow, which is flashy anyway.

William 1, The Dutch 'King of Canals', oversaw the digging of many navigation canals, most crucially the Zuid-Willemsvaart Canal in 1825 on what is now Belgian territory. With a view to diverting Meuse navigation from Liege to Antwerp on the Scheldt, the Meuse–Scheldt Junction Canal from was created in 1819. After Belgian independence in 1839, the Dutch presented the Belgians with the bill for canals affecting Belgian territory. This did not stop them extending the Zuid-Willemsvaart to create a twin shipping route from Liege to Antwerp, necessitating barrages upstream of Maastricht. In 1843, they diverted Meuse water into the Canal de la Campine, a canal connecting Antwerp with the Belgian east through the Zuid-Willemsvaart. A 1845 Convention allowed for the construction of a Liege-to-Maastricht canal; the treaty abolished intakes on Belgium territory and replaced them by an intake on Dutch territory. Belgium forwent navigation rights on the frontier section and the Netherlands could add intake to the Julianakanaal for navigation, but also gave Belgium extra water. The Belgians however also started to divert water for irrigating the Campany region, which increased the extremes of the already volatile Meuse river regime while drainage water caused floods on Dutch territory. A canal link between the Meuse and the German Moselle was never completed. After the Bocholt-Herentals Canal was dug in 1845 and Liege-Maastricht Canal in 1850 for drainage and irrigation, the Meuse became almost unnavigable due to all the Belgian diversions and the Dutch filed a complaint in 1851 that Dutch shipping interests were negatively affected. In 1862, the Dutch government stated interventions in transboundary rivers should not impinge negatively on co-riparians; making the Dutch–Belgian Meuse one of the earliest examples of the *do no harm* principle (Salman & Uprety, 2002).

The extension of the canal system fed by the Meuse caused the need for more water in times of drought, which a Treaty concluded in 1863 sought to realise. The terms of the treaty however did not reflect new Dutch and Belgian diversion works (Bouman, 1996). The issue continued to exercise the Dutch. The rise of coal mining in the early twentieth century made the Dutch again look for canalisation potential.

For any intervention, the Belgians as riparians needed to cooperate, which the Antwerp port authorities were loath to do even when Belgian industrialists joined Limburg industrial associations lobbying. Now an upstream country, Belgium had the power of obstruction on the Meuse by arresting the stream or flooding downstream areas. However, the Netherlands continued to control Belgium's outlet from Antwerp to the sea to the river Scheldt. From around 1900, the Meuse and Scheldt issues have been politically linked and in 1920–1926 were among the key political challenges between the two countries. After the Dutch had first invested in Rhine improvements, the Netherlands enacted the Law on Canalising the Meuse planning a canal from Ternaaien to Borgharen in 1915. The Belgians were too destroyed by the First World War to be able to object much. In 1921, the Dutch constructed the Juliana channel for barge traffic. The Belgians responded by digging their own channel, the Albertkanaal. This led to a stalemate. In 1936 the matter was brought before the International Court of Justice, which adjudicated in 1937, rejecting claims from both countries, thus solving nothing (Bouman, 1996; McIntyre, 2007).[5]

The Second World War put both Dutch and Belgian governments in exile in London. The joint suffering created sympathy for post-war collaboration, leading to the Benelux Economic Union in 1958. This seemed to spillover into better hydro-political relations on the Meuse when in 1961 a treaty was concluded to improve navigation by deepening both the entrance of the Juliana channel and the Albert channel. A 1963 Treaty explicitly linked the Scheldt to Meuse, requiring Belgium to compensate Netherlands for water flowing from Dutch territory to Belgium through the Rhine–Scheldt link. The only way would be to take extra water from the Meuse. A novelty was the introduction of water *quality* to the terms of the treaty. However, it was not as straightforward as it seemed. This will be explored below.

3.3. *1960s: Linking Water Quality with Quantity*

In part thanks to the Rhine/Scheldt connection and further interventions such as the Zandhoek sluice, the port of Antwerp continued to flourish in the twentieth century. The river Scheldt's right bank being full, development of the left bank was contemplated, not only straightening a sharp bend at Bath but also digging a new canal to serve the left-bank harbour facilities, the Baalhoek canal. This canal would largely be on Dutch territory. As the Dutch had no interest in helping the Belgians, they linked permission to three issues for which they sought upstream 'remediation of various negative externalities directed downstream' (Dombrowsky, 2009; Pham Do, Dinar, & McKinney, 2011): improvement on quality of Scheldt and Meuse water and guarantees of stabilised water quantity influx on the river Meuse (Meijerink, 2008).

The three draft 'water conventions' of 1975 resulting from negotiations started in 1963 ensured Belgium a minimum discharge of 50 m^3/s at all times, while Belgium was to construct large storage reservoirs upstream to cushion flash flood waves. The treaties however were never signed because of internal Belgian conflict. Upstream Wallonia (Francophone Belgium) felt it was forced to invest in clean-up measures and storage reservoirs but derived no benefits. The

Walloons refused to bear the brunt of the remedial measures the Dutch sought to secure. This led to internal stalemate within Belgium.

When in 1983 the Belgian government wanted to deepen the navigation channel of the West Scheldt, the Dutch linked this to the other issues (the bend at Bath had now disappeared from the agenda). The Belgians pushed for a lenient quality standard, while the Walloons wished to involve the French, whereupon which the Dutch suspended negotiations.

The UN-ECE transboundary watercourses (Helsinki) convention of 1992 however forced the involvement of France as upstream riparian to arrive at a comprehensive basin agreement. While UN-ECE provided a separate space to address the water-quality issues, the federalisation of Belgium made it possible to unlink the Flemish–Dutch talks over quantity issues from the multilateral issues. It enabled Flanders to break the impasse, negotiate as a separate unit, joining the deepening issue with the guarantee of a minimum flow while the international railway connection also became part of the package. This the Belgians now apparently see as a tactical error (Meijerink q. by Dombrowsky, 2009), the sort of 'excessive linkage' mentioned above. This raises the question: How many linkages are too much?

Belgian federalisation however also complicated negotiations on infrastructure measures. The quantity issue was initially solved with reasonable ease in early 1994 and France, the newly federalised Belgian regions and the Netherlands also made quick progress in negotiating conventions for the protection of the Meuse and Scheldt. But the Netherlands made a bilateral agreement on water issues conditional on progress on high-speed transboundary rail connection, *HSL*. In response, the Flanders government made its agreement to the multilateral Scheldt and Meuse conventions of 1994, conditional on Dutch agreement on the Scheldt deepening programme (Meijerink, 1999, p. 174, Gupta, van der Wurff, & Junne, 1995, pp. 121–122).

When this issue was finally ironed out, the other treaties could be signed. The Convention on the flow of the river Meuse proided for a water-saving scheme for times of low flow and established a working group for the regulation of the Meuse flow, which produced an Action Plan in 1998, if only in the most general terms; the convention for the protection of the Meuse and Scheldt led to the establishment of an International Commission on the Protection of the Meuse (ICMP) and one for the Scheldt (ICPS). The first Meuse Action Plan had been discussed at a Meuse Ministerial Conference. The city (region) of Brussels is a party to those meetings; while not a riparian, it depends on the river for its drinking water.

The International Meuse Commission (IMC) in 2000 observed a negative environmental impact of river intervention for navigation and hydropower and the need to integrated quantity (flood and drought) and quality issues. The IMC now has a flood working group, but find it difficult to make space for non-governmental organisations (Santbergen, 2012).

3.4. *Unilateral Action: The Border Meuse Project*

While progress was made between countries, domestic river policies were not integrated across borders and borders provinces started to look inward (Soeters, 1993). This was especially clear in flood policies.

As the river changed its course in the past millennia, the Meuse left a large fen in Central Limburg and as the earth crust rose, it caused the river to carve out new valleys, the older floodplain became what are now river terraces. As a consequence, the river is deeper than the hinterland, unlike the West Netherlands, so that floods do not cause damage beyond the area immediately bordering the river. Risk of flash floods however remains, intensified by channelisation. To reduce that risk, a Meuse re-naturation programme in Limburg was devised, the biggest current Dutch river project. The idea was to release the progressively normalised river from its straitjacket and thus create space for the river to meander (Helmer, Overmars, & Litjens, 1991). The relatively untamed French river Allier was the model for a natural river.

As river rehabilitation projects do not easily find a budget, the plan was to defray the costs of river rehabilitation from gravel sales from the Border Meuse. In Spring and early Summer, melted water from the Swiss glaciers upstream is transported down the river, eroding gravelly material which came down the Meuse as sediment and deposited in the lowest stretch. Gravel extraction however created ugly holes and nuisance. A plan to end gravel extraction in the 1990s was linked with the Border Meuse river restoration scheme.

The Meuse floods of late 1993 and early 1995 however opened a window of opportunity when the river Meuse rose to unexpectedly high levels. In 1994, a State Advisory Commission, called after its Chairman Boertien, recommended a flood defence programme that included Belgium. In 1995, the Meuse flooded again, causing considerable economic losses in both countries.

Cooperation between Holland and Belgium however was marginal, and 'in the border regions of Belgium and the Netherlands, warnings were given according to different gauge zero levels, with the result that predicted flood levels differed substantially across the border' (van Hassel & van Lindt, 1998, p. 75).

Emergency legislation was passed at breakneck speed in the Netherlands, and in this 'securitised' context promises from Cabinet Ministers pledging security from flooding by 2005 brought a strong popular legitimacy base to the Dutch Border Meuse plan. Widening the River would provide a higher degree of safety from floods as well, a unique emergency law in 1995 was rushed through both Chambers of Parliaments and Council of State, which evaluates every law ex-ante for legal implications. A programme of emergency flood defences (hard defences for the most populated areas, river widening elsewhere) was implemented in 'securitised' mode, meaning cost–benefit considerations, environmental impact assessments, and formal participation were swept aside in light of survival considerations, while expropriation of 'eminent domain' needed for flood defence was made possible.

When the securitised period ended in 1997 as the special law expired, a project organisation was created under the auspices of the national Waterways Agency, combining the Border Meuse with the Sandy Meuse stretch downstream as well as the Julianakanaal alongside. Finally, in line with the earlier advisory report, the plan also involved interventions on three Belgian locations, a sign of delinkage from security politics.

Cross-border protest: Everything looked promising in 1997. However between 1998 and 2001, there was radio silence on the part of the Meuse project initiators, breeding discontent and distrust. A multitude of grassroots platforms were up in arms against aspects of the envisaged Meuse Works. In 2001, when the plan was finally presented, local as well as conservationist organisations exploded, and provincial authorities felt compelled to take over by organising informal consultation for an alternative. This brought more structural dialogue with citizen platforms who, in 2003, joined hands to form the Bewoners Overleg Maaswerken (BOM). While BOM appreciated the open communication line, they did not like where the plans were going, and filed a case against the closed-shop project execution with the Dutch, then with European anti-trust authorities. For this they forged an unlikely strategic alliance with Belgian gravel companies who felt unduly excluded from tendering ('counterlinkage'). The province of Limburg had applied for dispensation of the trust rules with the Dutch anti-trust regulator NMa in 1991, arguing that the complexity of the project warranted this specific consortium rather than open tendering. The permit was especially important to the Dutch gravel extraction company, Panheel, as it saw its economic survival at stake, and sought to make use of the so-called right to self-realisation, which allows land owners to realise flood protection works on their land rather than be expropriated in the interest of eminent domain. This allowed land owners to oversee and implement flood protection measures, if paid from their own pockets.

A final salient point was the sudden disappearance from the plan of the Belgian bottleneck locations.[6] In the EIA hearings, nature conservationists had requested both banks to be treated in an integrated way. These locations were non-essential to achieving the required safety level on the Dutch side.[7] Moreover, Provincial authorities markedly noted that they did not want to insist on these locations as they basically wanted to provide safety 'on our own strength' *(in principe op eigen kracht)*, suggesting a reluctance to be dependent on the Belgians. In the final Maaswerken draft, therefore, the Belgian part had been scrapped with a view to cutting costs, to ensure budget neutrality, and saving time. Ultimately, the Dutch parties involved arrived at an agreement, under which only polluted clay had to be tendered. In 2005, the administrative agreement was signed and the Maaswerken started, if in heavily reduced form. But first some issues needed to be ironed out with the Belgian neighbours. The Dutch Maaswerken however was not designed to accommodate a 3000 m^3/s peak flow; for that, similar measures would be needed on the Belgian side (Teisman, 1995). The Belgians however are not always reciprocating naturalised river banks, and work by different time windows.[8]

Unilateral action in Belgium: In fairness cooperation also was not an issue for the Belgian federal government, which had never tabled flood defence since 1994. The Flemish authorities, who treated flood defence as a minor issue[9], planned on strengthening 45 km of their Common Meuse dikes between Lanaken and Maaseik (the Maasdijkenplan), protect an area subsided due to mining, and enhance environmental values in the project Living Border Meuse (Levende Grensmaas).

The Flemish had just adapted their regional strategic planning to incorporate their river works in 2001 when they learned the Dutch had scrapped the Belgian locations from their programme. However the Dutch had not counted on a Flemish veto due to reduced groundwater levels, which encumbered meeting the terms of the European Bird and Habitat Directive. It was agreed to counteract this by creating a gravel ridge on the Dutch side at Meers.

In view of the impending Dutch Border Meuse intervention, the Dutch government sent an emergency letter cautioning the Belgians in no uncertain terms, upon which a representative of the Belgian province of Limburg, also a member of the Dutch/Flemish committee, asked to implement the Belgian plans.[10]

The Flemish authorities for their part have been tempted to link their cooperation to the Grensmaas project with Dutch lenience on Belgian plans for reviving the disused Iron Rhine railway, which they claim a right to run in part through a Dutch Site of Special Scenic Interest, De Meinweg. In 1991, the Belgian government resolved to reinstate the freight transport link from Antwerp to the Ruhr industrial area in Germany. A 2000 report by Alterra, Wageningen, concluded the reinstatement of the railway line would be 'disastrous' (De Limburger, 2000). A diplomatic stalemate could have set the project back by years. The Permanent Court of Arbitration ruled in favour of Belgium (Djeffal, 2011).

The Belgians could easily have frustrated progress on the Maas works by making their cooperation in turn conditional on progress on the river Scheldt.[11] These threats however so far have not been concretised, and we can conclude that the Meuse has escaped political fallout from linkage politics (Warner, 2012).

Meanwhile it became clear to the Dutch that envisaged measures might not be climate-proof. The 3000 m^3/s ceiling might be broken in future if climate scenarios predicting higher variability were to come true. Integrated Maas Explorations (Integrale Maas Verkenningen, IVM) held with a range of Dutch stakeholders started in 2001, to safeguard Meuse safety for the long run beyond 2015, are premised on the idea that Belgium and France need to do their bit, to safeguard flood safety in case of a potential 4600 m³/s discharge (Wesselink, 2007).[12] To attain this ambitious goal, the co-operation of the Belgians would be badly needed. Sitting in on these exploratory scenario meetings (IVM2), Wesselink observed that while Belgian interests were noted in several of the intervention alternatives, the Dutch water authorities found it unnecessary to involve Belgians in these consultations nor take their flood safety intervention into account (Wesselink, Vriend de, Barneveld, Krol, & Bijker, 2009). While the steering committee of both IVM1 and IVM2 insisted on an integrated River basin approach, that is, including France and Belgium, the Dutch water agency did not assent.[13]

The Flemish river bank and branches of the Meuse were kept outside the scope, even on the map. How integrated could the explorations be?, as an alderman noted. The water agency claimed this was a 'parallel track' to be elaborated separately. Moreover the ongoing Belgian river rehabilitation measures were not taken into account, either. Geographic integration was clearly not what the commissioner had in mind (Wesselink, 2007). The Dutch continue to plan without taking upstream riparians into account. Not only the Delta Plan of 2008 but also its regional elaboration excluded Belgian interventions.[14] The latest developments however suggest some pragmatic cooperation.

3.5. *Doesn't Territory Matter Anymore?*

In 2006, as the Dutch had finally managed to conclude their domestic Meuse negotiation, a joint Dutch–Flemish project started to improve protection of residents between Eijsden and Maasmechelen as well as improve environmental values on the river banks. This flowed from the *Thalweg* issue: fixed at 1978 coordinates, the renaturated river's midpoint would shift as a result of the current interventions on both sides of the border. As a result, some 200 ha were to change hands as the Grensmaas progressed, making its shifts much more dynamic. This also impinges on rights to gravelling and proceeds thereof, and compensation for changes. In line with the late 1990s agreement, the Dutch province of Limburg has indeed freely transferred the nature park, Hochter Bampd, to Flanders in Summer 2008 to consolidate the transboundary project Levende Grensmaas, between Lanaken end Maasmechelen. The 35.4 ha reserve is on Flemish territory but historically owned by the Dutch Province of Limburg.[15] It forms part of a transboundary nature reserve of several hundreds of hectares Rivierpark Maasvallei, Natuurpark Maasvallei.

Most works were realised on Flemish territory and taken care of in terms of permiting, contracting, and implementation by the Belgians, but paid for by Dutch funding (EUR5 million) in light of the joint benefits. The river widening and floodplain excavation intervention near Lanaken is expected to have a positive effect on the city of Maastricht on its parishes, Itteren and Borgharen. The transboundary programme came under the auspices of the bilateral Flemish/Dutch Meuse Commission (VNBM),[16] established in 2005. In September 2013, the Dutch and Belgian Public Works Ministers were happy enough signed a new declaration of intention for similar implementation on the north part of the Border Meuse.

At the regional level, transboundary cooperation appears to be improving, though not without its problems.[17] The Action Plan against Meuse River Flooding envisages augmentation of flow capacity, installation of retention basins, greater community awareness of flooding, and a reliable flood warning system. VNBM has integrated the originally isolated Flemish and

Dutch gauging stations for water quantity and quality and ecology into transboundary stations. They also make sure the Dutch and Flemish use the same data and models for research and river modelling.

Despite the technical integration however, a Dutch water board interviewee interviewed by the author in 2012 saw his Belgian neighbours as being '30 years behind', wedded to hard defences, pouring concrete, and building high dikes. Cooperation has only been starting of late, focussed on technical exchange, and still there is need to chase up the water-level information. It has to go through many prediction models for each of the branches.[18] The Belgians can handle up to 3000 m³/s. The Belgian management of the upstream power plants at Ivoz and Lixhe[19] can lead to abrupt fluctuations in the river discharge (Achtersloot, 2003).[20]

4. Discussion

This section takes up the issue of linkage in different schools of thought in international (hydro-political) relations, with respect to regional integration. It ends with observations on the politics of linkage. I will not only focus on expressly declared linkages but also those analytically identifiable as such (Fahey, 2014).

4.1. *Application of the Framework*

While the Rhine is often held up by multilateral aid organisations to other conflictive basin riparians in Africa and Latin America and elsewhere,[21] it is easily forgotten that the Rhine Treaty itself took 40 years and several major environmental and political crises to emerge. Despite obvious interdependence in the hydropolitical security complex, the Rhine and Meuse watershed are not necessarily a water security community, that is, an area where violent conflict has become almost unthinkable (Deutsch, Burrell, Kann, & Lee, 1957). Clearly cooperation cannot be forced or facilitated just like that.

The present study has applied integration to the Meuse. The above indicates that political issue-linkage is neither easily separated from physical integration (interlinkage) and dissipation (through diversion canals), nor from territorial integration and dissolution. There is a fledgling literature however claiming that conflict and cooperation (Mirumachi & Allan, 2007), integration and fragmentation (Puntigliano & Briceno-Ruiz, 2013) do not have to exclude each other but can be complementary political forces for resilience despite crisis (Dabene, 2009). And while 'issue-linkage' has been employed, notably by Turkey in its hydropolitics with Syria on the shared Euphrates/Tigris, to achieve distributive ends in the discursive interaction over Turkey's Ilisu dam on the Tigris (Daoudy, 2009), there is no treaty on this river basin, but rather a situation of 'no war, no peace'.

This is useful to keep in mind in light of the tortuous road to integration on the river Meuse. We have seen a gradual extension of issues, from navigation-only to water quality and environmental regeneration. At the same time, we saw the gradual inclusion of more countries and non-state actors, which indeed made it possible to create basin-wide agreement for the Meuse. This was intensified by an increased interlinkage of physical water streams, which in principle can bring joint benefits (Sadoff & Grey, 2002) but also put pressure on the minimum supply of sufficient water in low-flow periods on the Meuse.

Given that the Border Meuse is a natural boundary between two sovereign countries, Dutch interventions easily have an impact across the border. Moreover upstream effects of downstream intervention are also always possible as water levels are pushed up there because of the intervention. It is therefore notable how the Maaswerken project mostly preferred *not* to rely on Belgian co-operation,[22] apart from some cooperation in the International Meuse Commission after the 1995 floods, which affected Holland, Belgium, and France.This changed somewhat in 2001, when the Treaty of Liege was signed by the Netherlands, Belgium, Flanders, Wallonia, and Luxembourg.

This delay in coordinated action is counterintuitive given that after the successful agreement of 1994, both countries started working on a programme river interventions on the Maas, and both were involved in tortuous non-river negotiations that risked spilling over into the river management arena. Structural linkages (such as links with the cut-throat competition between the two ports) as well as incidental linkage and 'counterlinkage' in many instances complicated things. This finding brings a nuance to received wisdom about linkage politics. On the basis of the above history, I would like to modify the drivers given in Table 1 with respect to linkage. Linkage may be absent or destructive rather than constructive for river management integration (columns B1 and B2). As shown in Table 3, which sketches Meuse linkage politics over time, these non-positive forms may coexist with 'positive' linkages, conducive to further cohesion. Ultimately, over time all these linkages impinge on the formation of a particular transboundary water regime. Conflict and cooperation may occur at the same time (Mirumachi & Allan, 2007; Warner & van Buuren, 2009) and may not necessarily be negative.

Linkage opens up the black box of conflict and cooperation, but also makes things more complex. The problems anticipated in wilfully coupling the Meuse and Scheldt seem to confirm the finding that when issue linkage becomes a complex Christmas tree (Warner et al., 2010), the resulting complexity may not only cause delays, but also result in 'conflicts, rigidity, or even stagnation' in which case delinking may be more appropriate (Brouwer, 2013). Communication is crucial in conditions of physical complexity. The physical complexity of the interlinked river Meuse, combined with cumbersome communication between Belgium and Netherlands, makes it harder for the Dutch to be proactive in flood management.

The above approaches help explain state behaviour, causing non-linear regime formation and integration on the Meuse. The linear trajectory from anarchy via selected regime to a security community assumed in much literature in a sense has happened. If we follow Lowi's (1993) Realist position that a downstream hegemon is necessary for cooperation, the Scheldt and Meuse qualify, with the Dutch in charge. A mostly coercive power until 1839, seeking to curb the Belgian harbour economically (Type A, 'anarchy'), the Netherlands later became a more benign hegemonic power, but clearly wary of sharing sovereignty, except when other European powers occupied Dutch territory (Spain, France, and Germany). Even at this stage (B, some regimes), the Belgians however have experienced the Netherlands as a dominant actor throughout, a shadow of the past (Sebastian, 2009) that continues to make itself felt (Warner & Zawahri, 2012). This is strongly related to the overriding economic security concern: economic survival of the ports. While cooperation in the International Meuse Commission has moved beyond a single-sector focus on shipping (type A non-integrated), it has long backgrounded flooding (Santbergen, 2012), in so doing falling short of Integrated Water Management (Phase C). After a securitised non-engagement in the 1990s, floods were desecuritised, and the latest cooperative flood defence initiative, if small, suggests in raised mutual awareness and relegation to river governance to 'low politics'.

The Meuse basin, physically and often politically interlinked with the Scheldt, is conspicuous for non-cooperation and conflict for much of its long history. Vries, Leibenath, Korcelli-Olejniczak, and Knippschild (2009) has noted that despite over 50 years of European integration, water can still easily become a diplomatic issue. The Meuse–Scheldt system are cases in point.

Table 3. Issue (non)linkage and counterlinkage and context over time through the Meuse's hdyropolitical history

Driver date	A. Conflict, power-driven or securitised non-linkage (−)	B1 Non-integration interest-driven counterlinkage or alliance (−)	B2 Integration interest-driven Cooperative linkage (+)	C Context-driven (+ and −)
1581–1815	Closing of Antwerp access, England supports this (*active, material, obstructive, tight*)	Belgium sides with Habsburgs		Dutch revolt against Hapsburgs breaks unified republic (−, *vertical*)
1815–1840	Canals in NL and B (*active, material, looser*)	Canals		Belgian (struggle for) independence, Limburg split (−, *vertical*)
1840–1915		Scheldt interests prevail over Meuse (*loose*)	• Limburg-Flemish joint Meuse lobby (*loose*) • Agreements on main channel without taking branches into account (*loose, nondurable*)	
1915–1940	Unilateral action NL, B			1915 B, F in World War, NL neutral (−)
1940 1960	Mutual claims NL and B, arbitrage, no solution			
1960–1980			Dutch–Flemish quality–quantity deal, externality shifted onto Wallonia (*loose, nondurable*)	• 1944 World War 2 B/NL govts in exile • 1951 Creation of Benelux, European Community (*loose*, +)

(*Continued*)

Table 3. Continued

Driver date	A. Conflict, power-driven or securitised non-linkage (−)	B1 Non-integration interest-driven counterlinkage or alliance (−)	B2 Integration interest-driven Cooperative linkage (+)	C Context-driven (+ and −)
1980s				Rhine treaty NL-D-F-CH after Sandoz pollution incident (+)
1990s	Non-coordinated river interventions after 1993/95 Meuse floods despite material transboundary effects	Counterlinkage with rail domain HSL, Iron Rhine (*obstructive, horiz.*)	• Meuse plan includes transboundary measures • River basin treaties (*constructive, durable*)	• UNECE (+) • Meuse Floods in NL, B, F (+) • Federalisation of Belgium (+)
1998–today	1998–2002 Meu(se programmes planned in isolation (*passive securitised non-linkage*)	2002 Crossboundary counterlinkage citizens NL & gravellers against Maaswerken (*obstructive, vertical*)	• 2005 Transboundary flood plan for Common Meuse • 2011 Territory returned to Belgium for nature de	• 2000 European Water Framework Directive (+) • 2001 Treaty of Liege

Notes: Numbers refer to category of linkage in text. +, enabling integration; −, complicating integration.

The Scheldt became a diplomatic incident as recently as 2008, and was only really resolved in 2012–2013. The Rhine likewise experienced an incident in 2010 over the so-called Crack Decision at the Haringvliet inlet, needed to allow salmon migration up the Rhine. The Meuse was not subject to direct linkage in either of these issues, and the change of Thalweg due to interventions, leading to a transfer of territory, did not bring the diplomatic tension over sovereignty some expected (Osinga, 1997). Conversely, Huisman, de Jong, and Wieriks, (1999) have argued that the Dutch leadership on the Rhine treaty has helped the Dutch take the lead in the OSPAR convention on reducing pollution of the high sea.

One factor recently hampering cooperation appears to have been the fear of floods and thus, human security. While the Meuse area is above sea level and as a consequence leaves more space to escape an oncoming flash flood than the Dutch lowlands, the province of Limburg has, largely for reasons of political expediency, managed to integrate the undiked Meuse in the national floods plan, making it a national interest and responsibility, in so doing securitising it (Wesselink, Warner, & Kok, 2012). The securitisation of Meuse flooding unified and centralised Dutch flood policy and discourse, but also promotes a sovereignty-driven approach rather than cooperation with Belgium. On the Meuse, 'inflexibility at the national level constrained social learning at the basin level' (Mostert et al., 2007). In this respect it is bemusing that the Dutch have not suffered human losses to floods on the Meuse and Scheldt except accidents during evacuation (in 1995 and 2012), while the Belgians have (both in 2010 and 2012).

The question whether water relations have promoted regional integration (*Context overspill,* Lindemann, 2008), suggested in regime literature, clearly merits an answer in the negative. While Europe can clearly be considered a security community, and the Netherlands was one of its founder members, integration has not significantly been promoted by water relations. For the Dutch and Belgian case, Meijerink (1999, 2008) however notes there is more to the context on regime formation on the Meuse than meets the eye. The creation of the Benelux and the European Community created an 'asecurity' zone, but does this context explain developments on the Meuse? Not quite. Meijerink notes that:

> the drafting of the UN-ECE Convention and the EU Water Framework Directive, may at first sight be interpreted as contextual developments. A better look at the processes that have been going on in these venues, however, shows that some negotiators involved in the negotiations over the Scheldt and Meuse water conventions were involved in the negotiations on the UN-ECE convention as well, and that these negotiators exploited the venue (. . .) to attract supporters to the image of international water management issues as being issues that have to be solved collectively by all basin states. Moreover, the same governmental bureaus which were involved in the negotiations on the Scheldt and Meuse water conventions, were involved in the preparatory work for the EU Water Framework Directive. In that process, representatives of (..) the Netherlands, along with other protagonists of the river basin approach, strongly put forward the concepts of integrated water management and river basin management. (Meijerink, 1999, p. 132)

There was thus an interplay between the formation of contextual institutions and the development in the Meuse and Scheldt river system. Linkages are complex!

4.2. *Linkage or Non-linkage?*

In the liberal reasoning, the drive for joint benefit may offset balance power asymmetry and promote regimes. This seems somewhat to be the case for the Meuse, though we have seen the story with regard to linkage is more complex. There is some evidence to suggest *issue linkage* can be really fruitful in promoting water agreements. As noted, however there are some ambiguities.

When we refer to issue-linkage, we usually mean active *(bargaining, discursive) linkages*. Water quantity and quality issues were linked from the 1960s, as well as railway transport, especially in the 1990s (Iron Rhine and HSL).

Material linkages have increased dramatically along with river linkages. Basin governance is naturalised, but in fact 'incorporates human alterations (dams, canals, and reservoirs, for example) into environmental regions' (Biro, 2007). These physical modifications however have tended to create nuisance (e.g. to shipping) and lawsuits rather than cooperation. Security has been conceived of as national security, preventing *political linkages* across borders for much of the river's history. The broadening of the actor base becoming more closely linked with the river management, whether for or against, has increased political linkages.

While Lindemann (2008) appears to paint cost incentives (issue linkage) as having a positive impact on regime creation, issue linkage can also leads to stalemate, which indeed it did for a long time on the Meuse and Scheldt. As Peeters (2009) notes, issue linkage may look attractive for its potential for generating package deals but is also hard to control. In the case of the high speed train line (HSL), linkage politics seemed successful. Meijerink has noted that it does not make sense to link issues where the other party does not benefit or will never agree. Issue linkage brought considerable delays.

The most obvious context for the Meuse has been the Scheldt, and this structural, non-material linkage has rarely been a positive one. Meuse and Scheldt negotiations could not be disentangled for over a century, are strongly related to economic competition between the ports Rotterdam and Antwerp, and seen as crucial to national interest (economic *securitisation*). While there are no such major ports on the Meuse, its physical and political links to the Scheldt have beset progress on this. While this ultimately turned out positive for the Scheldt, one may wonder if this was also the case for the Meuse.

Provocatively speaking, would cooperation on the Meuse not be more successful *without* issue linkage? On the Meuse, non-cooperation stands out. The Dutch and Belgians have conducted their own projects on the shared border with only a modicum of coordination, and a refusal to be dependent on the other for their security (physical securitisation: refusal of structural linkage). While the Border (or Common) Meuse and its tributaries have flooded regularly, threatening human security, the 1993 and 1995 floods impressed on both riparians the need to improve their flood protection infrastructure, but they largely went it alone. On this basis I have referred (Warner, 2012) to Belgian–Dutch Meuse relations as 'conspicuous boundary non-spanning': it was decided not to seek an integrated approach despite obvious opportunities.

In retrospect I may have been too pessimistic. Once river management was desecuritised, in 2005 the Dutch agreed to pay Belgians to undertake interventions that would have a positive impact on flood defence for the Dutch. There was a notable lack of concern for territory, the Dutch leaving 200 ha to Belgium to enable the Meuse works and the Dutch even paying for interventions on the Belgian side. Despite being bad neighbours on the rivers Rhine (non-compliance with closing the Haringvliet to enable salmon passage (Keessen, Hamer, Van Rijswick, Helena, & Wiering, 2012) and Scheldt (dispute with Belgium over depoldering the Hedwige polder on the Dutch–Belgian border) from 2008 to 2011, the Netherlands did not default on the river Meuse. This conceptual breakthrough, crossing boundaries in multiple ways, is a small but symbolic advance that may herald good things for the future.

Disclosure Statement

No potential conflict of interest was reported by the author.

Notes

1   A so-called neo-neo-consensus in International Relations however developed in the 1980s (Waever, 1997), where neo-Realists and liberal neo-institutionalists found common ground in understanding the formation and change of international cooperation in the absence of a global 'orderer'.

2   A 'boundary spanner' (after Williams, 2010) 'deliberately enters into multiple games, knowing that a losing position at one board may be compensated by a better situation at another board' (Warner & Zawahri, 2012).

3   Passive leverage: exerted just by existing. Active leverage: pressure is exerted.

4   Costs should, in this context, not be limited to economic transaction costs. As Keohane (1987) notes: without international regimes linking clusters of issues to one another, side payments (read: bribes) and strategic linkages would be difficult to arrange in world politics. In the absence of a price system of favours, institutional barriers would hinder the construction of bargains.

5   The Dutch wanted control not only over the Dutch Meuse intakes but also its Belgian intakes, while the Belgians objected to lowering the water level at the Borgharen barrage (at Maastricht). Neither claim was justified, as long as the quantity and level of water in the feeder channel did not change. The treaty only governed the left bank, so alternations at the right bank, such as the Juliana Channel, did not come under the treaty, and so the court ruled. This gave countries considerable leeway in altering the flow in their territory without taking into account the other party's interest. In its ruling of 1937, the International Court of Justice, The Hague, also noted that equality of riparians not only pertains to navigational but also non-navigational uses of the river.

6   'Nederland heeft schuld aan getreuzel Grensmaasproject', *Het Belang van Limburg*, 3 March 2003, http://www.hbvl.be/cnt/oid233922/archief-nederland-heeft-schuld-aan-getreuzel-grensmaasproject (last consulted 10 December 2015).

7   *Startnotitie m.e.r-Grensmaas* (2002) Online: http://www.commissiemer.nl/docs/mer/p12/p1270/1270-02sn.pdf (Last consulted 10 December 2015).

8   As marl excavation on the Belgian side has given rise to a 7m drop very close to the river (Mijnverzakkingsgebied), this is not always feasible. Interview V. Coenen, Maaswerken stakeholder €manager. 2005.

9   *op de laatste vergadering stuurde het zijn kat.*

10   http://www.hbvl.be/Archief/guid/provincie-wil-herinrichting-grensmaas-zelf-realiseren.aspx?artikel=c5a3b2d3-88e0-4e24-97eb-eac7943b550e

11   The deal on a planned fareway deepening on the estuary of that river basin in exchange for nature development fell apart when the Dutch farmers successfully engaged the Dutch prime minister to undo the flooding of a scenic Zeeland area, the Hedwigepolder. Predictably the Belgian co-riparians on the Scheldt felt the Dutch were breaching their trust, and threatened to (re)link the Scheldt to other issues—the so-called 'Iron Rhine' railway, Dutch traffic on the Antwerp circular road, even the mussel trade, but also the management of the river Meuse (Meijerink, 2008, Warner & van Buuren, 2009).

12   'Alweer nieuw plan voor beveiliging Maas', *De Limburger*, 19 December 2001.

13   Wesselink (2007) reports a heated debate in the working sessions: 'it makes no sense to proceed unless the exercise becomes an Integrated Basin Exploration for the Meuse'. Other participants arguing that negotiations within the Me use catchment had so far not yielded any concrete results and the Netherlands could not afford to be dependent and wait for others to take action.

14   *De uitgevoerde Vlaamse maatregelen langs de Maas worden niet meegenomen in de referentie-situatie* (Implemented Flemish measures along the Meuse are not taken into account in the reference situation).

15   Werken levende grensmaas gaan dit najaar van start, *Het Belang van Limburg,* 2 July 2008. Online: http://www.hbvl.be/cnt/aid731199/werken-levende-grensmaas-gaan-dit-najaar-van-start (last consulted 10 December 2015)

16   This body involves, on the Flemish side, the Waterways agency (nv De Scheepvaart), the Public Works ministry, the Nature and Forests Agency, the Flemish Province of Limburg, the Ministry of Environment, Nature, and Energy and Flanders International. On the Dutch side, the Waterways agency Rijkswaterstaat, the Ministry of Infrastructure and Environment, the Ministry of Economic Affairs, the Roer en Overmaas water management board, and the Dutch Province of Limburg.

17   This is reportedly very different from cooperation with Germany on the Roer (Rur)—the Germans immediately inform the Dutch on the detail of the level of the artificial lake on the German side. (interview)

18   Such as the Ourthe, which feeds Liege, six hours before a flood peak reaches the city of Maastricht.

19   The HEPP Ivoz-Ramet has been in use since 1954 and for more than 300 days each year uses the total discharge of the River Meuse.

20   Interviewee from water board: 'Cooperation between the Netherlands and Belgium is still in its infancy. They can hold the water for longer, or they can let it pass through. They have a hydropower plant near the border, which also

handles 2–300 m³/s, while the Albertkanaal takes 200 m³/s. Wallonia does not take Limburg into account at all. But 200 m³ more or less makes the difference between flooding the road (between Borgharen and Itteren, two Maastricht parishes, JW) or not. There is now the Amice project [for better coordination]

21 The ongoing World Bank CIWA project is one example. Online: http://blogs.worldbank.org/water/peoplemove/strengthening-cooperative-management-and-development-of-international-waters-in-africa

22 To be fair there was some consideration of cross-border effects—to avoid that the bunds around Roosteren would increase flood risk on the Belgian side at maaseik, it was agreed that the Meuse would be widened on the Dutch side near Roosteren.

23 Strategy is about setting the objective(s), the what-for; tactics is about the way the objectives are to be reached, the how (Goldratt, Goldratt, & Abramov, 2002). Tactical linkage (Sung, 2009) is the way outsiders persuade, with sufficient leverage, strong and hegemonic insiders to accept the agenda the weaker actors wanted to put in place. Finally, what would be the opportunity cost of non-linkage?

24 This pertains to both intention and outcome: Does the linkage lead to a deal? Is it ignored, does it bring alienation, is it overburdened?

References

Achtersloot, R. (2003). *MER Grensmaas 2003—Achtergronddocument 1—Rivierkunde*. Commissioned by Rijkswaterstaat, De Maaswerken. Project 9M4711.A0 Royal Haskoning/10064 Meander Advies en Onderzoek.

Berkes, F. (2002). Cross-scale institutional linkages: Perspectives from the bottom up. In E. Ostrom, T. Dietz, N. Dolsak, P. C. Stern, S. Stonich, & E. U. Weber (Eds.), *The drama of the commons* (pp. 293–322). Washington, DC: National Academy Press.

Bernauer, T. (2002). Explaining success and failure in international river management. *Aquatic Sciences, 64*(1), 1–19.

Biro, A. (2007). Water politics and the construction of scale. *Studies in Political Economy, 80*, 9–30.

Bouman, N. (1996). A new regime for the Meuse. *Review of European Community and International Environmental Lam, 5*(2), 161–168.

Bow, B. J. (2009). *The politics of linkage: Power, interdependence, and ideas in Canada-US relations, Vancouver*. Toronto: UBC Press.

Brouwer, S. (2013). *Policy entrepreneurs and strategies for change* (PhD thesis). VU University.

Buntaine, M. T. (2007). *Regional integration, issue fragmentation, and cooperative environmental governance in the Lancang-Mekong river basin*. Retrieved from http://www.2007amsterdamconference.org/Downloads/07SummerSchool%20-%20Buntaine.pdf

Busuttil, S., Calleja, J., & Wiberg, H., (Eds.). (1994). *The search for peace in the Mediterranean region. Problems and prospects*. Valletta: Mireva.

Buzan, B., & Waever, O. (2003). *Regions and powers: The structure of international security*. Cambridge Studies in International Relations. Cambridge: Cambridge University Press.

Buzan, B., Waever, O., & de Wilde, J. (1998). *Security: A new framework*. Hemel Hampstead: Harvester Wheatsheaf.

Dabene, O. (2009). *The politics of regional integration. Theoretical and comparative explorations*. New York, NY: Palgrave McMillan.

Daoudy, M. (2009). Asymmetric power: Negotiating water in the Euphrates and Tigris. *International Negotiation, 14*, 359–389.

Davis, C. L. (2004). International institutions and issue linkage: Building support for agricultural trade liberalization. *American Political Science Review, 98*(1), 153–170.

De Limburger. (2000, November 12). Alterra: IJzeren Rijn ramp voor Meinweg. *De Limburger Regional Daily*.

Deutsch, K. W., Burrell, S. A., Kann, R. A., & Lee, Jr., M. (1957). *Political community and the North Atlantic area; international organization in the light of historical experience*. Princeton: Princeton University Press.

Djeffal, C. (2011). The Iron Rhine case: A treaty's journey from peace to sustainable development, *Zeitschrift für Ausländische Öffentliches Recht und Völkerrecht* 71. Retrieved from http://www.zaoerv.de/71_2011/71_2011_3_a_569_586.pdf (last consulted 10 December 2015)

Dombrowsky, I. (2009). *Benefit-sharing in transboundary water management through intra-water sector issue linkage?* Stockholm Water Week.

Duivenvoorden, A. (1997). *In de maas verdiept. Een regionaal geografische verkenning van bron tot monding*. Amsterdam: NIVON/NFI.

Fahey, C. (2014). *Linkages in the postwar Canada-US relationship*. Working Paper 5. Institute for Transnational Study of the Americas. Brock University and the University of Buffalo-the SUNY.

Goldratt, E., Goldratt, R., & Abramov. (2002). *Strategy and tactics*. Research paper. Retrieved from http://public.wsu.edu/~engrmgmt/holt/em534/Goldratt/Strategic-Tactic.html, last consulted 10 December 2015

Gupta, J., van der Wurff, R., & Junne, G. (1995). *International policies to address the greenhouse effect. An evaluation of international mechanisms to encourage developing country participation in global greenhouse gas control strategies, especially through the formulation of national programmes*. Amsterdam: Institute for Environmental Studies, Vrije Universiteit.

Haggard, S., & Simmons, B. A. (1987). Theories of international regimes. *International Organization, 41*(3), 491–517.

Hasenclever, A., Mayer, P., & Rittberger, V. (1996). Interests, power, knowledge: The Study of International Regimes. *Mershon International Studies Review, 40*(2), 177–228.

Hassel, H., & van Lindt, L. (1998). Flood management in Belgium. In U. Rosenthal, P. t Hart, & M. Bezuijen, (Eds.), *Flood response and crisis management in Western Europe: A comparative analysis* (pp. 57–101). Berlin: Springer Verlag.

Heikkala, T., Schlager, E., & Davis, M. W. (2011). The role of cross-scale institutional linkages in common pool resource management: Assessing interstate river compacts. *Poly Studies Journal, 39*(1), 121–145.

Helmer, W., Overmars, W., & Litjens, G. (1991). *Toekomst voor een Grindrivier*. Nijmegen: Stroming, Bureau voor landschapsontwikkeling in opdracht van Provincie Limburg.

Huisman, P. de Jong, J., & Wieriks, K. (1999). Transboundary cooperation in shared river basins: Experiences from the Rhine, Meuse and North Sea. *Water Policy, 1*(1–2), 83–98.

Jägerskog, A. (2013). *Transboundary water management—why it is important and why it needs to be developed*. Retrieved from http://www.watergovernance.org/documents/WGF/Reports/TWM-why-it-is-important.pdf

Keessen, A. M. Hamer, J., Van Rijswick, M., Helena, F. M. W., & Wiering, M. (2012). The concept of resilience from a normative perspective: Examples from dutch adaptation strategies. *Ecology and Society, 18*(2), 45. Retrieved from http://www.ecologyandsociety.org/vol18/iss2/art45/

Keohane, R. O. (1987). The demand for international regime. *International Organisation, 36*(2), 325–355.

Keohane, R. O., & Nye, J. (1977). *Power and interdependence. World politics in transition*. Boston, MA: Little, Brown.

Kistin Keller, E. (2010). *Cooperation in transboundary waters* (PhD dissertation), Oxford University.

Legro, J. W., & Moravcsik, A. (1999). Is anybody still a realist? *International Security* 24(2): 5–55. Retrieved December 10, 2015, from http://www.princeton.edu/~amoravcs/library/anybody.pdf

LeMarquand, D. G. (1977). *International rivers: The politics of cooperation*. Vancouver: Westwater Research Centre, University of British Columbia.

Levitsky, S., & Way, L. A. (2010). *Competitive authoritarianism: Hybrid regimes after the cold War*. Cambridge, MA: Cambridge University Press.

Lindemann, S. (2008). Understanding water regime formation—A research framework with lessons from Europe, *Global Environmental Politics, 8*(4), 117–140.

Lohmann, S. (1997). *Linkage* politics. *Journal of Conflict Resolution, 41*(1), 38–67.

Lowi. (1993). *Water and power. The politics of scarce resources in the Jordan River basin.* Cambridge: Cambridge University Press.

Lukes, S. L. (1974 [2005]). *Power, A radical view.* London: McMillan.

Maaswerken, De. (2004), *Terugblik* 2004.

McIntyre, O. (2007). *Environmental protection of international watercourses under international Law.* Aldershot: Ashgate.

Meijerink, S. (2008). Explaining continuity and change in international policies: Issue linkage, venue change, and learning on policies for the river Scheldt estuary, 1967–2005. *Environment & Planning A, 40*(4), 848–866.

Meijerink, S. V. (1999). *Conflict and cooperation on the Scheldt river basin: A case study of decision making on international Scheldt issues between 1967 and 1997.* Dordrecht: Kluwer Academic.

Meissner, R., & Turton, A. R. (2003). The hydrosocial contract theory and the Lesotho highlands water project. *Water Policy, 5*, 115–126.

Mirumachi, N., & Allan, J. A. (2007). *Revisiting transboundary water governance: Power, conflict, cooperation and the political economy.* CAIWA conference paper 2007. Retrieved from http://www.newater.uni-osnabrueck.de/caiwa/data/papers%20session/F3/CAIWA-FullPaper-MirumachiAllan25Oct07submitted2.pdf

Mitrany, D. (1965). The prospect of European integration: Federal or functional. *Journal of Common Market Studies, 4*(2), 119–149.

Mitrany, M. (1948). The functional approach to world organization. *International Affairs (Royal Institute of International Affairs), 24*(3), 350–363.

Mostert, E., Pahl-Wostl, C., Rees, Y., Searle, B., Tábara, D., & Tippet, J. (2007). Social learning in European river basin management; barriers and fostering mechanisms from 10 river basins. *Ecology and Society, 12*(1), 19. Retrieved from http://www.ecologyandsociety.org/vol12/iss1/art19/

Oelsner, A. (2005). *(De)securitisation theory and regional peace: Some theoretical reflections and a case study on the way to stable peace.* EUI (European University Institute Robert Schumann Centre of Advanced Studies) Working paper, RSCAS No. 2005/27. Retrieved from http://cadmus.eui.eu/bitstream/handle/1814/3249/2005_27.pdf?sequence=1

Osinga, J. (1997). *De grens in de Maas. Een onderzoek naar de grens en de Nederlands-Belgische regelingen inzake de rivier de Maas* [The border in the River Meuse. An investigation into the border and the Dutch-Belgium regulations regarding the river Meuse]. Maastricht: Wetenschapswinkel, Universiteit van Maastricht.

Peeters, M. (2009). The joint governance of transboundary river basins. Some observations on the rule of law. In Faure & Soon Yong (Eds.), *China and transboundary liability* (pp. 192–224). London: E Elgar.

Perrow, C. (1984 [1999]). *Normal accidents living with high-risk technologies.* Princeton, NJ: Princeton University Press.

Pham, Do, K. H., Dinar, A., & McKinney, D. (2011). *Can issue linkage help mitigate externalities and enhance cooperation.* MPRA Paper No. 37408, Munchen. Retrieved from http://mpra.ub.uni-muenchen.de/37408/1/MPRA_paper_37408.pdf (last consulted 10 December 2015)

Puntigliano, A. R., & Briceno-Ruiz, J. (eds.). (2013). *Resilience of regionalism in Latin America and the Caribbean. Development and autonomy.* New York, NY: Palgrave MacMillan.

Putnam, D. (1988). Diplomacy and domestic politics: The logic of two-level games. *International Organization, 42*(3), 427–446.

Puchala, D. J., & Hopkins, R. F. (1983). International regimes: Lessons from inductive analysis. In Stephen D. Krasner (Ed.), *International Regimes* (pp. 61–90). Ithaca: Cornell University Press.

Rosenthal, U., Hart, P 't., & Bezuijen, M. (1998). *Flood response and crisis management in Western Europe. A comparative analysis.* New York, NY: Springer.

Sadoff, C. W., & Grey, D. (2002). Beyond the river: The benefits of cooperation on international rivers. *Water Policy, 4*(5), 389–403.

Salman, S. M. A., & Uprety, K. (2002). *Conflict and cooperation on South Asia's International rivers: A legal perspective.* Dordrecht: Kluwer International Law, 14–16.

Santbergen, L. L. P. A. (2012). *Ambiguous aspirations in the Meuse theatre* (PhD thesis). Nijmegen: Radboud University.

Schulz, M. (1995). Turkey, Syria and Iraq: A hydropolitical security complex. In L. Ohlsson, (ed.), *Hydropolitics: Conflicts over water as a development constraint* (pp. 107–113). London: Zed Books.

Sebastian, A. G. (2009). *Transboundary water politics: Conflict, cooperation, and shadows of the past in the Okavango and orange river basins of Southern Africa.* PhD Dissertation, University of Maryland.

Soeters, J. L. (1993). Managing Euregional networks. *Organization Studies, 14*(5), 639–656.

Stokke, O. S. (1997). 'Regimes as governance systems'. In O. R. Young (ed.), *Global governance: Drawing insights from the environmental experience* (pp. 27–63). Cambridge: MIT Press.

Sung, K-Y. (2009). *Success and failure in dealing with North Korea: Has issue-linkage worked?* Paper presented at the BISA Conference 2009.

Teisman, Geert R. (1995). Het project grensmaas. *Bestuurskunde, 4*(8), 370–380.

Tol, R., & Langen, N. (2000). A concise history of Dutch river floods. *Climatic Change, 46*(3), 357–369.

Turton, A. R. (2002). *The political aspects of institutional development in the water sector: South Africa and its international river basins* (D.Phil. thesis). Department of Political Science, University of Pretoria, South Africa.

Turton, A. R., & Meissner, R. (2003). The hydropolitical contract and the lesotho highlands project. *Water Policy, 5*(2), 115–126.

Voza, D., Vuković, M., Carlson, L., & Djordjević, D. B. (2012). International water conflict and cooperation: The role of power relations among riparians. *International Journal of Humanities and Social Science, 2*(11), 5.

Vries, J. de, Leibenath, M., Korcelli-Olejniczak, E., & Knippschild, R. (2009). Cross-border cooperation in the Rhine-Scheldt basin. The long road to institution building.

Waever, O. (1996). The rise and fall of the inter-paradigm debate. In S. Smith, K. Booth, & M. Zalewski (Eds.), *International theory: Positivism and beyond* (pp. 149–183). Cambridge: Cambridge University Press.

Warner, J. (2011). *Three lenses on water War, peace and hegemonic struggle on the nile. Journal of social sustainability.* Geneva: Inderscience.

Warner, J. (2012). Framing and linking space for the Grensmaas: Opportunities and limitations to boundary spanning in Dutch River management. In J. Edelenbos, N. Bressers, & P. Scholten (Eds.), *Water governance as connective capacity* (pp. 89–109). Aldershot, UK: Ashgate.

Warner, J., & van Buuren, A. (2009). Multi-stakeholder learning and fighting on the river Scheldt. *International Negotiation, 14*(2), 419–440.

Warner, J., Lulofs, K., & Bressers, H. (2010). The fine art of boundary spanning: Making space for water in the East Netherlands. *Water Alternatives, 3*(1), 137–153.

Warner, J., & Zawahri, N. (2012). Hegemony and asymmetry: Multiple-chessboard games on transboundary rivers. *International Environmental Agreements: Politics, Law and Economics, 12*(3), 215–229.

Wendt, A. (1999). *Social theory of international politics.* Cambridge, MA: Cambridge University Press.

Wesselink, A., Warner, J., & Kok, M. (2012). You gain some funding, you lose some freedom: Hegemony in flood protection in the Netherlands. *Environmental Science and Policy, 30*, 113–125.

Wesselink, A. J. (2007). *Integraal waterbeheer: de verweving van expertise en belangen* (PhD dissertation). Twente University, Rnschede.

Wesselink, A. M., Vriend de, H., Barneveld, H., Krol, M., & Bijker, W. (2009). Hydrology and hydraulics expertise in participatory processes for climate change adaptation in the Dutch Meuse. *Water Science and Technology, 60*(3), 407–413.

Wester, Ph., & Molle, F. (2009). *River basin trajectories.* Colombo: International Water Management Institute (IWMI).

Williams, P. (2010, February). *Special agents: The nature and role of boundary spanners.* Paper to the ESRC Research Seminar Series: 'Collaborative Futures: New Insights from Intra and Inter-Sectoral Collaborations', University of Birmingham.

Wit, M. J. M. de, Peeters, H. A., Gastaud, P. H., Dewil, P., Maeghe, K., & Baumgart, J. (2007). Floods in the Meuse basin: Event descriptions and an international view on ongoing measures. *Internation Journal of River Basin Management, 5*(4), 279–292.

Young, O. R., Ed. (1999). *The effectiveness of international environmental regimes: Causal connections and behavioral mechanisms.* Cambridge, MA: MIT Press.

Escaping the Border, Debordering the Nature: Protected Areas, Participatory Management, and Environmental Security in Northern Patagonia (i.e. Chile and Argentina)

BASTIEN SEPÚLVEDA & SYLVAIN GUYOT

ABSTRACT *This paper focuses on the management of protected areas in transboundary contexts and centres on the contemporary evolution of the border between Chile and Argentina in Northern Patagonia, which is a region that has witnessed the creation of numerous protected areas that are currently claimed by Mapuche organisations and communities as part of their customary territory. In response to these claims, both states have progressively integrated Mapuche communities into the management of protected areas through specific agreements. Many of these protected areas have also been included in a Transboundary Biosphere Reserve proposal for UNESCO. A new environmental governance model that includes the protection of indigenous peoples' rights is under construction not only along but also across the border between Chile and Argentina. Therefore, we discuss how participatory management could be viewed as a tool for redefining borders by linking environmental security in protected areas to human security in Mapuche communities. The article seeks to understand the role of environmental governance in shaping and/or overcoming political boundaries, and analyse how strategic mobilisations of the environment can advance the achievement of competing territorial projects led by different actors in different periods.*

1. Introduction

Prior to the rise of the modern nation states in Latin America, the Andes region did not include rigid and fixed political borders. On the contrary, it was once a meeting point for economic trade and sociocultural exchanges between different indigenous groups.[1] The Andean region, therefore, was a strategic link that articulated vast indigenous territories, such as the Mapuche territory in Northern Patagonia, which is one of the most renowned examples of aboriginal trans-Andean territorialities (Pinto, 1996). Historically, the genesis of Chilean and Argentinean states did not imply the immediate dismemberment of Mapuche territory. However, geopolitical stakes led Chile to engage in processes of territorial expansion that finally broke previous spatial configurations towards the late nineteenth century. The incorporation of Northern Patagonia, which was intended to increase the land available for agricultural colonisation, was particularly violent. In fact, the military campaigns of 'Araucania's Pacification' on the Chilean side (1861–1881) and 'Conquest of the Desert' on the Argentinean side (1869–1888) were simultaneously achieved. Consequently, the Mapuche territory became politically disintegrated, which forced neighbouring communities to enclose themselves and evolve into different national realities.

Despite evident interest in the historical trans-Andean (dis)articulations (Bello, 2011; Pinto, 1996) and imposition of national narratives on both sides of the border (Bandieri, 2009; Donoso, 2008), little attention has been paid to contemporary indigenous mobilities and territorial (re)configurations across the Andes (Gundermann, González, & De Ruyt, 2009; Molina, 2009). However, an increasing body of literature highlights the transnational dimension of the contemporary Mapuche movement and its political discourses (Aylwin, 2009; Boccara, 2006a, 2006b; Kradolfer, 2010). In addition, numerous scholars have discussed the naturalisation of the Andes as a political border (Bandieri, 2001; Núñez, 2013; Núñez, Sánchez, & Arenas, 2013) and the role of nature conservation in the border-making process (Aagesen, 2000; Navarro Floria & Delrio, 2011; Núñez, Matossian, & Vejsbjerg, 2012). Indeed, this literature is useful for understanding the importance of protected areas in both the construction and articulation of the border and in providing concrete substance to national narratives.

Nonetheless, little attention has been given to the environmental dimension of contemporary cross-border dynamics, even though a Transboundary Biosphere Reserve (TBR) proposal for UNESCO has been discussed in recent years by representatives from Chile and Argentina (Adriazola & Araya, 2007). Mapuche organisations and communities currently claim many protected areas included in this proposal. In response to these claims, Chile and Argentina are implementing experimental co-management agreements that attempt to integrate Mapuche communities at different levels of decision-making (Aylwin & Cuadra, 2011; García, 2011; Guyot, 2010; Miniconi & Guyot, 2010; Pérez Raventós & Biondo, 2003; Sepúlveda, 2012). Thus, a new environmental governance model appears to be under construction along and across the border between Chile and Argentina.

Can we consider participatory management as an effective tool for border redefinition? May this new form of environmental governance successfully link environmental security in protected areas to human security in Mapuche communities? How does it connect and/or reinforce indigenous territorial projects? To develop such a reflection, we draw on specific literature and primary data collected in the field between 2007 and 2012 through participant observation and informal interviews with local stakeholders and institutional representatives. The interest in nature conservation and environmental governance in transborder contexts is not new, and previous case studies from Europe (Fall, 2003) and Africa (Wolmer, 2003a) have highlighted the benefits and challenges of transboundary conservation. What distinguishes the selected case study from others is its coincidence with a customary territory reclaimed by a specific indigenous group. As such, the TBR proposal between Chile and Argentina would potentially link environmental security to human and cultural concerns and eventually address the resolution of indigenous rights, which is a specific body and extension of human rights characterised by its collective dimension (Anaya, 2004).

Considering that Chile decided not to participate in any regional structures before joining the Union of South American Nations (UNASUR) in 2008, we may wonder if an original form of regional integration could emerge from this specific configuration. Overall, Chile adopted the weakest form of political integration by signing specific bilateral free trade agreements, with nearly 50 agreements currently signed with different countries. Does this lack of formal integration represent a threat or an opportunity, and if so, to whom? May this context be favourable to the acknowledgement of indigenous rights through the implementation of the TBR proposal? At the crossroads of the borderland and environmental and indigenous studies, this paper seeks to understand the role of environmental governance in shaping and/or overcoming political boundaries. We analyse how strategic mobilisations of the environment can advance the achievement of competing territorial projects led by different actors in different periods.

Through the case of Northern Patagonia, we discuss the conceptual shift in the use of the environmental security paradigm from national to human security. We demonstrate that environmental security was initially absorbed by national security, fully participating in border demarcation and serving nation-state purposes (Part One). Based on wilderness enclosure, this former process created strong socio-spatial exclusions that primarily affected aboriginal communities located on either side of the border. Currently, indigenous territorial claims over protected areas contest this specific type of territorial domination and contribute to redirect environmental security from national to human security (Part Two). The last section discusses the ways in which the Chilean and Argentinian states address these claims and focuses on whether current environmental policies overcome, redraw, or strengthen the international border (Part Three).

2. Border, Protected Areas, and National Security in Northern Patagonia

The long-disputed border between Chile and Argentina was formally established in 1882 and then marginally modified in 1902 following arbitration by the British Crown. In this section, we analyse the role of protected areas in Chile and Argentina in consolidating, securing, and even articulating this political border in the first half of the twentieth century, which saw various nature reserves and national parks proclaimed to strip indigenous people of their lands while securing state's borderlands. We further argue that by serving nation-state purposes, this process has created spaces of social exclusion and territorial domination that have affected local Mapuche communities.

2.1. *Securing the Border: The Role of Protected Areas in Border Regions*

In Chile, the first protected areas were created on remnant parcels of land from major auctions through which the state established a significant number of Chilean and European settlers. Contrary to conventional wisdom, protected areas were not created to address the ecological impacts of uncontrolled forestry, especially in Araucania where thousands of hectares of native forests were burned to provide space for agricultural development. In fact, almost all of the protected areas created in this period were intended to reaffirm the state's sovereignty over resources that had to be preserved within a 'resourcist perspective' (Ramousse & Salin, 2007).

Nonetheless, a closer analysis reveals that most of these protected areas were located in the Andes to secure the border that separates Chile from Argentina in a period in which Northern Patagonia was more of an 'open frontier' than a strong political border (Bandieri, 2001; Pinto, 2011) (see Figure 1). In 1912, three protected areas—Alto Bío-Bío, Villarrica, and Llanquihue forest reserves—were established to fill the unused spaces along the international border. Further steps were then taken to consolidate the border when certain sectors of the former forest reserves gained autonomous status, producing the first national parks in Chile: the Vicente Pérez Rosales National Park, which was separated from the Llanquihue Forest Reserve in 1926, and the Villarrica National Park, which was separated from the Villarrica Forest Reserve in 1940 (Cabeza, 1988). In addition, the Puyehue National Park was created in 1941 along the border with the official purpose of promoting tourism.

The same process occurred in Argentina in which 'conservation, in its origin, was not simply associated with the border-making process but also entailed the manipulation of the environment towards an idealised image, constructed through tourism [. . .]' (Núñez et al., 2012, p. 54). This was particularly evident when the Southern National Park, the first protected area in Argentina, was established in 1922 (Navarro Floria & Delrio, 2011). With the creation of the National Park Administration (NPA) in 1934, it became the Nahuel Huapi National Park. As a prelude to the proclamation of many other 'border parks' in Argentina, the creation of the Nahuel Huapi National Park constituted one of the essential historical elements towards securing the territorial stability of the region bordering Chile (Miniconi & Guyot, 2010). Between 1934 and 1954, six other national 'border parks' were officially proclaimed along the international borders of Argentina: one along the Brazilian border in 1934 (Iguazú National Park), four along the Chilean border in 1937 (Lanín, Los Alerces, Perito Moreno, and Los Glaciares National Parks), and another along the Paraguayan border in 1951 (Río Pilcomayo National Park) (Núñez et al., 2012).

'Border parks', which are often so-called empty spaces, constitute real geopolitical reserves in the context of international border conflicts. In the 1960s, the emerging ecological movements pleaded for the enforcement of new environmental security principles within border protected areas because of their poor environmental management and landmine dispersal.

2.2. *Protected Areas and Indigenous Peoples: Spaces of Territorial Domination*

Following the Yellowstone National Park model, protected areas in Chile and Argentina were originally conceived and defined as spaces dedicated exclusively to nature conservation in which tourism was generally the sole human activity allowed. As a result, local and indigenous communities were deprived of a portion of their customary territories so that pristine landscapes could be conserved for tourists. At that time, environmental security was a national issue that was considered incompatible with human security. This context progressively changed in the second half of the twentieth century when Argentina and Chile adopted in 1970 and 1984, respectively, the International Union for Conservation of Nature (IUCN) categories as a reference framework (Núñez et al., 2012; Sepúlveda, 2012) (Table 1).

Contrary to strict conservation objectives defined for national parks (Cat. II) and natural monuments (Cat. III) that prevent any form of human intervention, the conservation objectives of national reserves (Cat. IV) are formally associated with local communities and stakeholders. In Chile, national reserves (formerly forest reserves) were officially defined as 'areas in which it is necessary to carefully conserve and use the natural resources, because of their susceptibility to degradation or relevance in safeguarding the community's well-being'.[2] Beyond this definition, the national reserves remained under the control of a strong hierarchical administration that

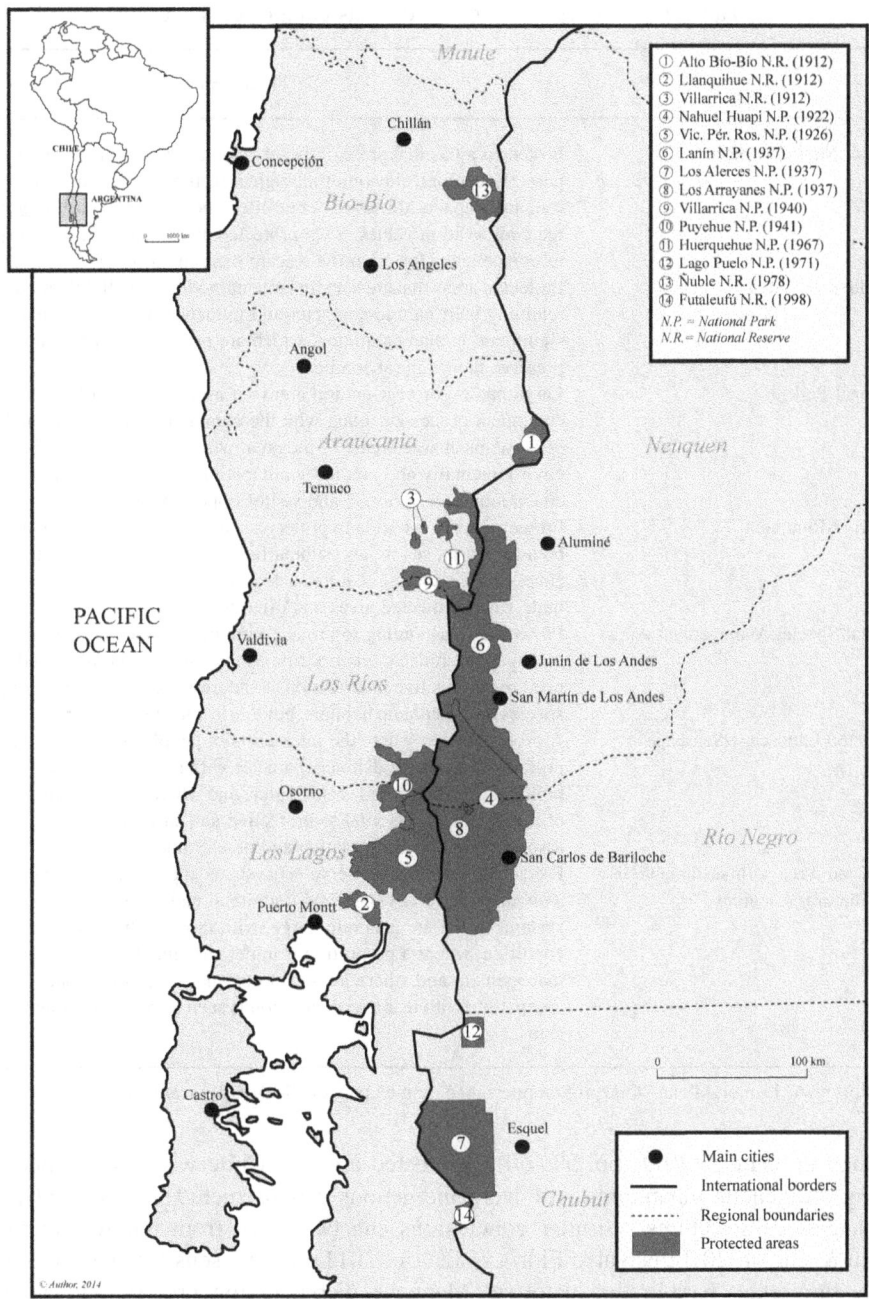

Figure 1. Border parks created in Northern Patagonia (i.e. Chile and Argentina). *Source*: Author.

excluded local communities from the decision-making process. Moreover, the management plan of the reserves was delegated to external experts hired by the *Corporación Nacional Forestal de Chile* (CONAF), which is the national agency in charge of forestry and parks. Most of these experts, however, are not necessarily aware of the local realities.

Table 1.   The IUCN protected areas categories system

Category	Definition
I a Strict Nature Reserve	Protected areas that are strictly set aside to protect biodiversity and also possibly geological/geomorphological features, where human visitation, use, and impacts are strictly controlled and limited to ensure protection of the conservation values. Such protected areas can serve as indispensable reference areas for scientific research and monitoring
b Wilderness Area	Protected areas that are usually large unmodified or slightly modified areas, retaining their natural character and influence, without permanent or significant human habitation, which are protected and managed so as to preserve their natural condition
II National Park	Large natural or near-natural areas set aside to protect large-scale ecological processes, along with the complement of species and ecosystems characteristic of the area, which also provide a foundation for environmentally and culturally compatible spiritual, scientific, educational, recreational, and visitor opportunities
III Natural Monument	Protected areas set aside to protect a specific natural monument, which can be a landform, sea mount, submarine cavern, geological feature such as a cave, or even a living feature such as an ancient grove. They are generally quite small protected areas and often have high visitor value
IV Habitat/Species Management Area	Protected areas aiming to protect particular species or habitats and management reflects this priority. Many category IV protected areas will need regular, active interventions to address the requirements of particular species or to maintain habitats, but this is not a requirement of the category
V Protected Landscape/Seascape	A protected area where the interaction of people and nature over time has produced an area of distinct character with significant ecological, biological, cultural, and scenic value, and where safeguarding the integrity of this interaction is vital to protecting and sustaining the area and its associated nature conservation and other values
VI Protected Area with sustainable use of natural resources	Protected areas that conserve ecosystems and habitats, together with associated cultural values and traditional natural resource management systems. They are generally large, with most of the area in a natural condition, where a proportion is under sustainable natural resource management and where low-level non-industrial use of natural resources compatible with nature conservation is seen as one of the main aims of the area

Source: http://www.iucn.org/about/work/programmes/gpap_home/gpap_quality/gpap_pacategories/.

As stated by García (2011, pp. 59–60), 'protected areas in Chile were established without consulting indigenous inhabitants, nor were indigenous groups included in formal protected area management structures'. Similar conclusions can be drawn from the Argentinean case (Miniconi & Guyot, 2010; Navarro Floria & Delrio, 2011). In this sense, protected areas have played an important part in the historical Mapuche dispossession process and appear as a form of territorial domination that has been pursued and reinforced over time in a context wherein the legal framework does not provide for genuine participation with local communities and stakeholders. Nonetheless, over the last two decades, the Mapuche people from Argentina and Chile have begun a mobilisation process at different scales to reclaim protected areas as part of their customary territories. These mobilisations not only attempt to redefine the allocation of natural resources for indigenous communities: participatory processes in which numerous indigenous people have engaged worldwide actually tend to be considered a first step towards the recognition and recovery of indigenous sovereignty.

3. Escaping the Border: From National to Human/Cultural Security

Indigenous voices in international fora stress the so-called sustainability of traditional practices and 'harmonious relationship' that indigenous peoples are supposed to maintain with natural resources. The rhetoric of sustainability tends to construct and convey representations of aboriginal peoples as innate ecologists or 'ecological natives' (Ulloa, 2005). Indigeneity therefore appears to have found an incontestable nature-based legitimacy that is actively mobilised within current territorial claims (Kent, 2008; Sepúlveda, 2012). In this respect, ecologism and indigeneity may be considered connected dimensions of a same discourse. In addition, this rhetoric allows indigenous peoples to engage in contemporary debates over the environment. By promoting a new form of environmental governance based on the principles of social participation and shared responsibilities, indigenous representatives have helped to move nature conservation from the paradigm of 'locked spaces' (Amelot & André-Lamat, 2009) to that of sustainable development. In this way, they have enriched the understanding of environmental security by attempting to highlight and incorporate its human, and even cultural, dimension.

3.1. *Indigenous Conservation Territories?*

In response to strong indigenous mobilisations in the international arena since the 1970s, almost every current indigenous rights instrument has asserted the importance of the relationship that aboriginal peoples maintain with the environment. The 169 ILO Convention, for example, highlighted in its Article 15 the 'rights of the peoples concerned to the natural resources pertaining to their lands shall be specially safeguarded. These rights include the right of these peoples to participate in the use, management and conservation of these resources'.[3] Accordingly, agreements and conventions related to biodiversity also emphasised the role of aboriginal peoples in the conservation of natural resources. The Convention on Biological Diversity, adopted in 1992 at the Río Earth Summit, stressed in its eighth article (letter j) the commitment of the parties to

> [...] respect, preserve and maintain knowledge, innovations and practices of indigenous and local communities embodying traditional lifestyles relevant for the conservation and sustainable use of biological diversity and promote their wider application with the approval and involvement of the holders of such knowledge, innovations and practices [...].[4]

Not surprisingly, aboriginal peoples are often the most prominent opponents to strict nature conservation processes and contest the social exclusion mechanisms associated with the implementation of protected areas. Actually, native leaders have participated for many years in international meetings related to the management of protected areas and have expressed their specific concerns. In this respect, the second edition of the Latin American Congress of National Parks and other Protected Areas conducted in 2007 in Bariloche, Argentina, was particularly relevant. On this occasion, the *Confederación Mapuche de Neuquén* (CMN) organised an alternative meeting in which more than 80 native leaders from different Latin American countries participated (Nahuel, 2008). Drawing on international declarations and conventions on indigenous rights, aboriginal representatives prepared a position paper in which they exhorted the members of the conference to encourage the creation of a new conservation category:

> [...] we conclude that it is time to evaluate the management categories that are currently recognized. The clearest demand is reflected in the need for the creation of Indigenous Conservation

Territories as a new management category under the control of our indigenous and afro organizations and regulated by indigenous laws as a tool for the control and administration of those territories. Indigenous peoples do not live within Protected Areas. Protected Areas are within indigenous territories, and thanks to our use models, work and traditional knowledge we can bequeath to the world what little is left for the future generations. (Nahuel, 2008, pp. 22–23)

This specific claim was reported and addressed in the official declaration of the conference in which the 'ecological dimension' of aboriginal practices was fully acknowledged. In this sense, the ability of indigenous peoples to fit the 'native factor' into the agenda of global nature conservation must be considered as one of the most relevant aspects of their political mobilisations in recent decades. However, native leaders do not only advocate for more environmental justice. The environmental issues contained in aboriginal discourses also refer to and articulate with strong territorial claims aiming to improve indigenous control over customary territories and regain indigenous sovereignty. In Northern Patagonia, such a process occurs simultaneously on both sides of and across the Chile–Argentina border.

3.2. *From Environmental Governance to (Transnational) Indigenous Sovereignty*

Many scholars have stressed the transnationalisation of the contemporary Mapuche political movement that allows aboriginal representatives from Argentina and Chile to share a common agenda in which the environment holds a central role (Aylwin, 2009; Boccara, 2006b; Kradolfer, 2010). The 2007 Bariloche meeting fully participated in this dynamic by incorporating protected areas within a huge territorial recovery process across the Southern Andes. Mapuche representatives from Argentina and Chile have met on multiple occasions in recent years and built common strategies that have strengthened the feeling of national belonging among people from both sides of the Andes. For instance, a national flag—the Wenu Foye—was created at the beginning of the 1990s. From an initiative of the *Consejo de Todas las Tierras* (CTT), which has been one of the most influential Mapuche organisations in Chile since the 1990s, this flag has been used on both sides of the border as a strong symbol of the Mapuche reunification process (Boccara, 2006b). Aboriginal communication agencies have also been a part of this process.[5] By simultaneously covering news of the Mapuche political struggles in Chile and Argentina, the aboriginal press is undoubtedly contributing to the renewal of Mapuche geographical imaginaries.

Similarly, many Mapuche intellectuals from Chile have engaged in a process of revising and reinterpreting official historical narratives in favour of acknowledging a 'Mapuche National History' (Mariman, Caniuqueo, Levil, & Millalen, 2006). Whilst contesting the arbitrariness of the political boundary drawn in 1882 and consequent imposition of the Argentinean and Chilean states in Northern Patagonia, Mapuche intellectuals highlight past aboriginal territorial unity and trans-Andean articulations. Anthropologist Boccara argued that 'the denaturalizing of national territorial divisions and borders is concomitant with a process of indigenous space renaturalizing' (Boccara, 2006a, §10). A note published in 2002 on the Mapuche International Link (MIL) website delved into the structure of this 'renaturalised space':

The Mapuche nation is situated in what is known as the Southern Cone of South America, in the area now occupied by the Argentinean and Chilean states. [...] The Mapuche nation is situated in its historical ancestral territory, the Wall-Mapu: Wall; universe, Mapu; land/territory. [...] The traditional political structure of the Mapuche people is reproduced within the structure of the Mapuche territorial entity. It is organised into four geographical regions or Meli wixan-mapu. Each wixan-mapu is made up of aylla rewe (eight districts) which, in turn, are made up of communities known as lof.[6]

In this statement, administrative districts, provinces, regions, and even states are not considered. Mapuche geographical discourse reveals and asserts the existence and legitimacy of a trans-Andean country with its own labels, jurisdictions, and administrative system. This country, which is referred to as *Wallmapu*, clearly appears to be what Agnew and Oslender (2013) call an 'overlapping territoriality', which has even been drawn on alternative maps wherein the Andes disappear as a political border (Boccara, 2006a). The upside down maps of the Mapuche historian Mariman illustrate this intention to reveal silenced indigenous geographies (see Figure 2).

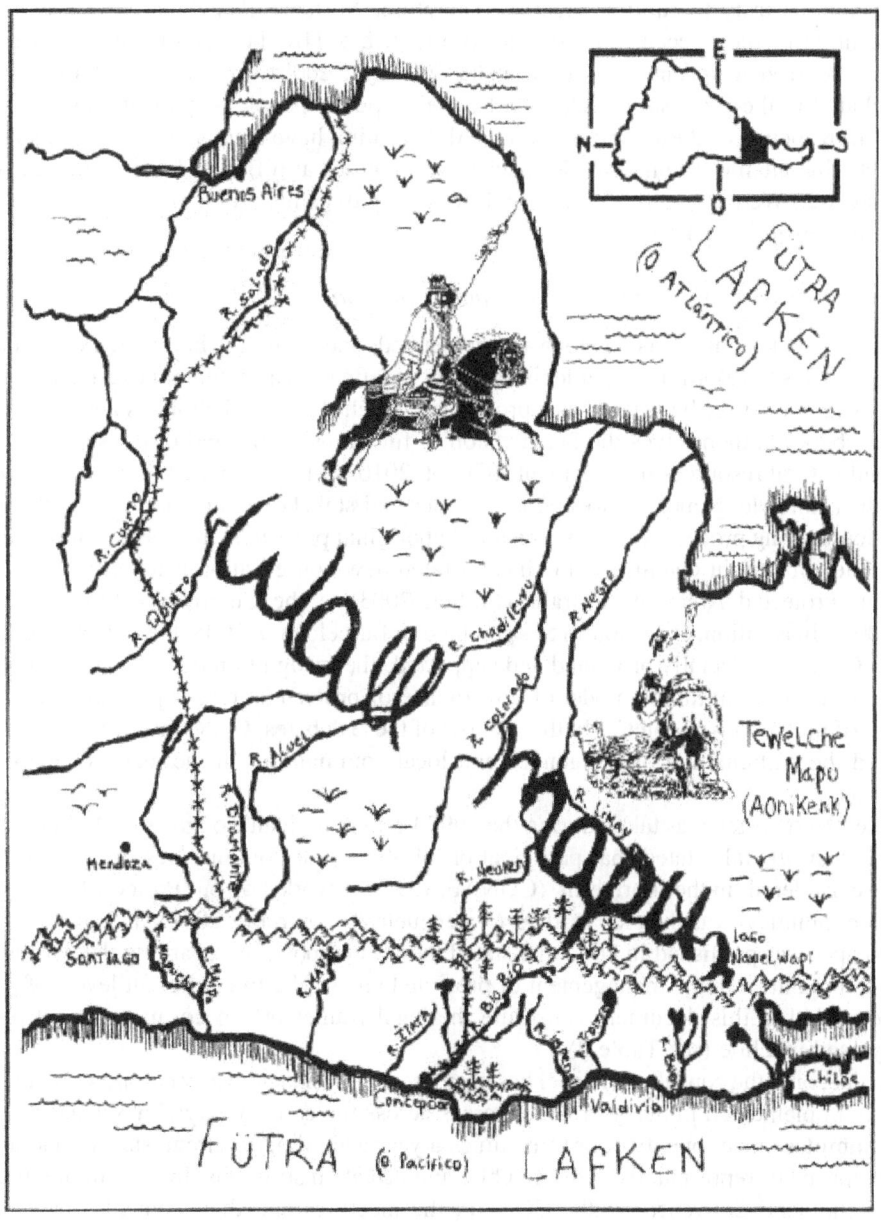

Figure 2. Mariman's upside down map of the Mapuche territory. *Source*: Mariman (2002, p. 54).

4. Debordering the Nature? Institutional Responses on Both Sides of the Border

The aboriginal report submitted to the 2007 Latin American Congress of National Parks stated a series of recommendations with respect to 'indigenous transboundary territories'. One of these recommendations was to 'enable mobility, exchange and integration amongst peoples; as well as experiences of coordination for the management of protected areas and indigenous territories (ecological and cultural corridors)' (Nahuel, 2008, p. 76). By explicitly connecting environmental issues with indigenous sovereignty, this specific topic clearly associated environmental security to cultural security. Therefore, Northern Patagonia can be considered a significant case study because a proposal to establish a TBR between Chile and Argentina has been discussed in recent years (Adriazola & Araya, 2007). In addition, many protected areas included in the proposal are claimed by the Mapuche people as part of their customary territory. In response to these claims, Chile and Argentina have implemented co-management agreements that attempt to integrate Mapuche communities at different levels of the decision-making process. We explore in this section how such a dynamic would articulate and respond to Mapuche territorial claims.

4.1. *Co-management Agreements in Chile and Argentina*

Participatory management is a long-recommended measure in international texts that encourages states to reform national legislations to redefine environmental governance. Participatory management may be defined as a process by which '[...] stakeholders negotiate, define and secure between themselves the fair division of functions, rights and responsibilities of territorial and natural resources management' (Guyot, 2010, p. 168). Applied to protected areas, it implies the acknowledgement of local communities and stakeholders by associating them with the decision-making process. Often as a result of aboriginal pressures, the participatory management policies of certain countries have incorporated new conservation categories, such as the Indigenous Protected Areas in Australia (Muller, 2003) or the Communal Reserves in Peru (Kent, 2008). In addition, at its Fourth Session held in Barcelona in 2008, the IUCN World Conservation Congress officially promoted and supported 'the recognition of Indigenous Conservation Territories as a legitimate model of governance of protected areas superimposed over the territories of indigenous peoples'.[7] In the shadow of these debates, Chile and Argentina recently considered the participation of indigenous and local communities in the management of protected areas.

In Chile, the first step was taken during the 1997 Latin American Congress on National Parks, where Chilean officials stated that participation of aboriginal communities in protected areas was to be considered 'in the short term' (CONAF, 1997). A workshop on 'Protected areas—Indigenous communities' was then led in 1999 (Valenzuela & Contreras, 2000), and an institutional working paper was published in 2002 that stated the official position regarding the participation of local communities in the management of protected areas. The five different levels of participation identified in this document constitute the legal framework of co-management policies currently used in Chile (see Table 2).

In practice, only the consultation level has been used, and local advisory committees have been effectively implemented in many protected areas across the country (Aylwin & Cuadra, 2011). These committees meet one, two, or three times a year according to circumstances and are formally composed of representatives from public and private institutions, local communities, and other stakeholders. Convened by the CONAF, the meetings and their contents are generally organised by an institutional representative. This hierarchical order has only been overridden

Table 2.   Official levels of co-management in Chile

Level		Modes of participation
1	Information	Communities are invited to assist only at informational meetings in a top-down manner
2	Consultation	Communities are invited to discuss specific projects or concerns in the framework of local advisory committees that partner the entire local population
3	Association	The community may join forces with the institution by signing an agreement to achieve specific goals
4	Collaboration	Collaboration occurs for institutional subsidised projects supporting community organisations
5	Integration	Integration of the community in management planning activities, including the defining of boundaries of protected areas

Source: Araya (2002).

on the occasion of association agreements, which are the third level of participation, being signed between CONAF and local communities to achieve specific goals. Time limitations and strict regulations apply to these agreements; therefore, they are not pertinent frameworks for redefining power relationships within protected areas. Nevertheless, indigenous Non Governmental Organisations (NGOs) from Chile perceive and promote association agreements as an interesting empowerment tool, and as a result of CTT negotiations with CONAF regional representatives, many association agreements have been signed in different protected areas of Araucania since 2000 (Aylwin & Cuadra, 2011).

In Argentina, an interest in participatory management publicly emerged when the CMN occupied the Lanín National Park administrative building in September 1999 to support the local Mapuche communities in their territorial claims of the protected area.[8] In response to these mobilisations, a workshop on Protected Indigenous Territory was organised in San Martín de los Andes in May 2000. This meeting resulted in new cooperation principles and a common declaration that validated a series of basic requirements on territory, co-management, and cultural–biological linkage. As a result of this process, a Co-management Committee was formalised in July 2000 between the NPA, CMN, and local communities. Contrary to the Chilean local advisory committees, which are strictly led by institutional representatives, the Lanín Co-management Committee is directed jointly by the park's intendant and a Mapuche representative (Pérez Raventós & Biondo, 2003).

In addition, the NPA board officially incorporated the principle of co-management in its 2001 Institutional Management Plan, which asserted that 'aboriginal peoples will have a key role in the development of the areas they inhabit through the co-management of these'. Accordingly, an Advisory Committee for Indigenous Policy was formally created in the NPA structure in September 2007. Its main objective consisted of looking after the different measures taken by the NPA regarding aboriginal territories. Following the path drawn by the CMN in the Lanín National Park, an Intercultural Co-management Committee was also established in April 2012 in the Nahuel Huapi National Park. Despite this progress, the CMN states that co-management is not sufficient to reach their goals of land, territory, and rights recognition, showing a lack of consideration of cultural security principles by the authorities (CMN, 2009). Since then, the CMN has been proactive in holding such discussions in international arenas to strengthen their position and legitimise their will at a local level considering the new nexus between environmental and cultural security.

4.2. *The Alto Bío-Bío National Reserve Experience in Chile*[9]

The Alto Bío-Bío National Reserve was established in 1912 in Lonquimay, a municipality located in the Andean section of the Araucania Region in the Mapuche-Pehuenche customary territory. The initial objective of this protected area was to secure the international border between Chile and Argentina. However, in 1929, the continuous influx of Chilean peasants into the sector forced the authorities to acknowledge the settlement of many families on more than 5000 ha within the boundaries of the reserve. As a result, the total area of the protected area was reduced to 31,000 ha. Strong land ownership conflicts remained, especially with the Mapuche-Pehuenche communities that had occupied this area long before the creation of the reserve. Since the early 1990s, the CONAF has received requests from many competing actors, indigenous and non-indigenous peasants, in search of lands for breeding activities during the summer season—known as *veranadas*. By accessing the same areas every year, livestock farmers tended to develop a sense of belonging and exclusivity, primarily based on the basic differentiation between *Colonos* (i.e. Chilean peasants) and the indigenous people. Thus, ethnic identity appeared over time as a type of spatial denominator.

In that context, in 2006, representatives from the different Mapuche-Pehuenche communities of Lonquimay created the *Consejo Pewenche de Lonquimay* or CPL, whose structure is essentially based on the strong kinship ties maintained over several generations among the members of the organisation. One of the main purposes of the CPL was to reclaim a large portion of the Alto Bío-Bío National Reserve, which had been 'illegally' created within the boundaries of the Pehuenche customary territory according to the organisation. The Pehuenche representatives were also integrated into, and influenced by, the CTT and could thus benefit from both its experience and fame. This clearly explains how, just a few months after the founding of their organisation, the Pehuenche leaders negotiated an association agreement with the CONAF.

Signed on 23 May 2007, this agreement was conceived as a basic, global framework that would be made more precise and enhanced by specific agreements with each indigenous community. From this perspective, in 2008, the Pehuenche representatives submitted to the CONAF a document in which they clearly defined three bounded areas for community *veranada* use and a fourth sector claimed collectively as 'sacred territory'. The latter, known as Cerro Bayo, is identified as an emblematic place where the last Mapuche warriors may have found refuge during the 'pacification' campaign. For this reason, and to put the agreement into practice, the members of the CPL eventually chose that place to celebrate a great *nguillatun*—the most important ceremony in the indigenous religious system.

However, the Pehuenche leaders did not seem to pay much attention to the opposition of the *Colonos*, who also had strong economic interests at Cerro Bayo. Joined by many Pehuenches with whom they shared the same devotion for the Pentecostal faith, the *Colonos* blocked the road leading to Cerro Bayo, which is an international highway between Chile and Argentina. For its part, the CONAF, under pressure from the local authorities such as the mayor of Lonquimay, eventually decided to withdraw its authorisation for the ceremony on the previously approved celebration day. Overwhelmed by the situation, the institution even allowed the park rangers to leave the reserve. This experience clearly shows the limitations of the CONAF prerogatives and action framework. The inability of the institution to understand and address the local political rivalries and the sociocultural dynamics led to a deadlock situation that temporarily broke the participatory process. Not sufficiently aware of the territorial stakes linked to the creation of the protected area, the institutional representatives, through their incorrect decisions and management, reinforced the existing interethnic and intra-ethnic divides.

4.3. The Creation of Biosphere Reserves Along (and Across?) the Border

As previously mentioned, many of the protected areas currently subjected to participatory management policies in Chile and Argentina have been included in a TBR proposal (Adriazola & Araya, 2007). The Biosphere Reserve Program, which was initiated in 1974 by UNESCO, had a strong influence in Northern Patagonia. A first Biosphere Reserve (BR) was declared in Chile in 1983: the Araucarias Biosphere Reserve was established in Northern Araucania along the border with Argentina. This primary entity was extended in 2010 and now covers a total area of 1,142,850 ha, which includes many protected areas. Argentina and Chile were also engaging in discussions regarding the creation of a TBR to transform Northern Patagonia into a shared territory. The establishment of two BRs on both sides of the border marked an initial and fundamental step of this process. The Bosques Templados Lluviosos de los Andes Australes Biosphere Reserve in Chile and the Andino Norpatagónica Biosphere Reserve in Argentina were both created in 2007 and cover 2,168,956 and 2,266,942 ha, respectively. Although the planned TBR has not been formally created, three existing BRs were joined as an ecological corridor that fully covers the international border between Chile and Argentina in Northern Patagonia. Thus, the Southern Andes appear as a de facto cluster of internationally adjoining protected areas (Figure 3).

According to Fall (2003), TBRs are part of a second generation of BRs that emerged in the post-Cold War context at a time when nature conservation was experiencing 'the return to planning on the scale of nature' (Fall, 2002, p. 249), paying more attention to the ecosystem approach and concepts such as bio or ecoregions (Wolmer, 2003a, 2003b).[10] There is a consensus in the literature on the increasing interest in and number of internationally adjoining protected areas, which increased from 136 complexes in 1997 to 169 in 2001 (Zbicz, 2003). We accordingly noted a multiplication of different formal labels that refer to this specific configuration of nature conservation. The IUCN, for example, officially refers to Transboundary Protected Areas (TBPAs), which include formal categories such as TBRs and are defined as

> areas of land and/or sea that straddle one or more boundaries between states, sub-national units such as provinces and regions, autonomous areas and/or areas beyond the limits of national sovereignty or jurisdiction, whose constituent parts are especially dedicated to the protection and maintenance of biological diversity, and of natural and associated cultural resources, and managed co-operatively through legal or other effective means. (Sandwith, Shine, Hamilton, & Sheppard, 2001, p. 3)

It must be added that a specific recommendation expressed by the IUCN is that TBPAs should be established 'where local communities and indigenous peoples in natural areas are linked across boundaries by shared ethnic or socio-cultural characteristics, traditions and practices' (Sandwith et al., 2001, p. 44). By linking conservation objectives to humanitarian concerns, TBPAs help redirect the environmental security paradigm from a national-based perspective to a more human and even cultural perspective. It thus appears as an appropriate figure to address indigenous sovereignty. As Wolmer (2003a, p. 264) argued, TBPAs

> [...] constitute a means of re-establishing cultural—as well as ecological—integrity, or achieving the 'cultural harmonisation' of divided ethnic groups. Removing the artificial natural boundaries dividing ethnic groups, it is hoped, will 're-establish historical links' and 'foster a cultural renaissance'.

Nation states are both assisted and pressured in these processes by other stakeholders, such as indigenous and environmental NGOs, with the former ensuring the respect for aboriginal rights and the latter in charge of the designation of protected perimeters, such as the WWF Global 200 initiative. The proposal for a TBR across the Southern Andes occurs in this

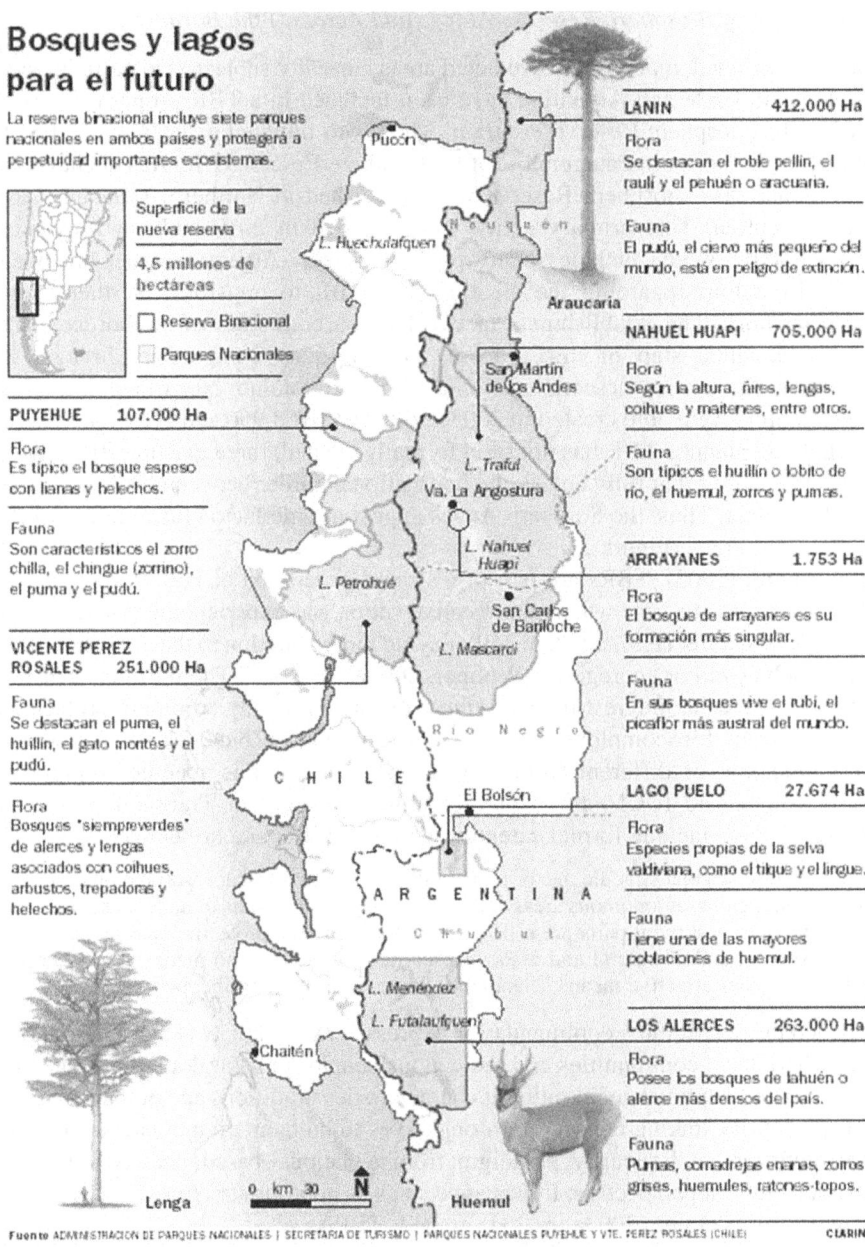

Figure 3. The TBR proposal between Chile and Argentina. *Source*: Navarro Floria (2008).

context. Many of the protected areas included in this proposal have been grouped together and identified by IUCN in two of the 169 complexes of internationally adjoining protected areas listed in 2001. All of them also belong to the Valdivian Temperate Forest Ecoregion identified by WWF through its Global 200 Program (see Figure 4).

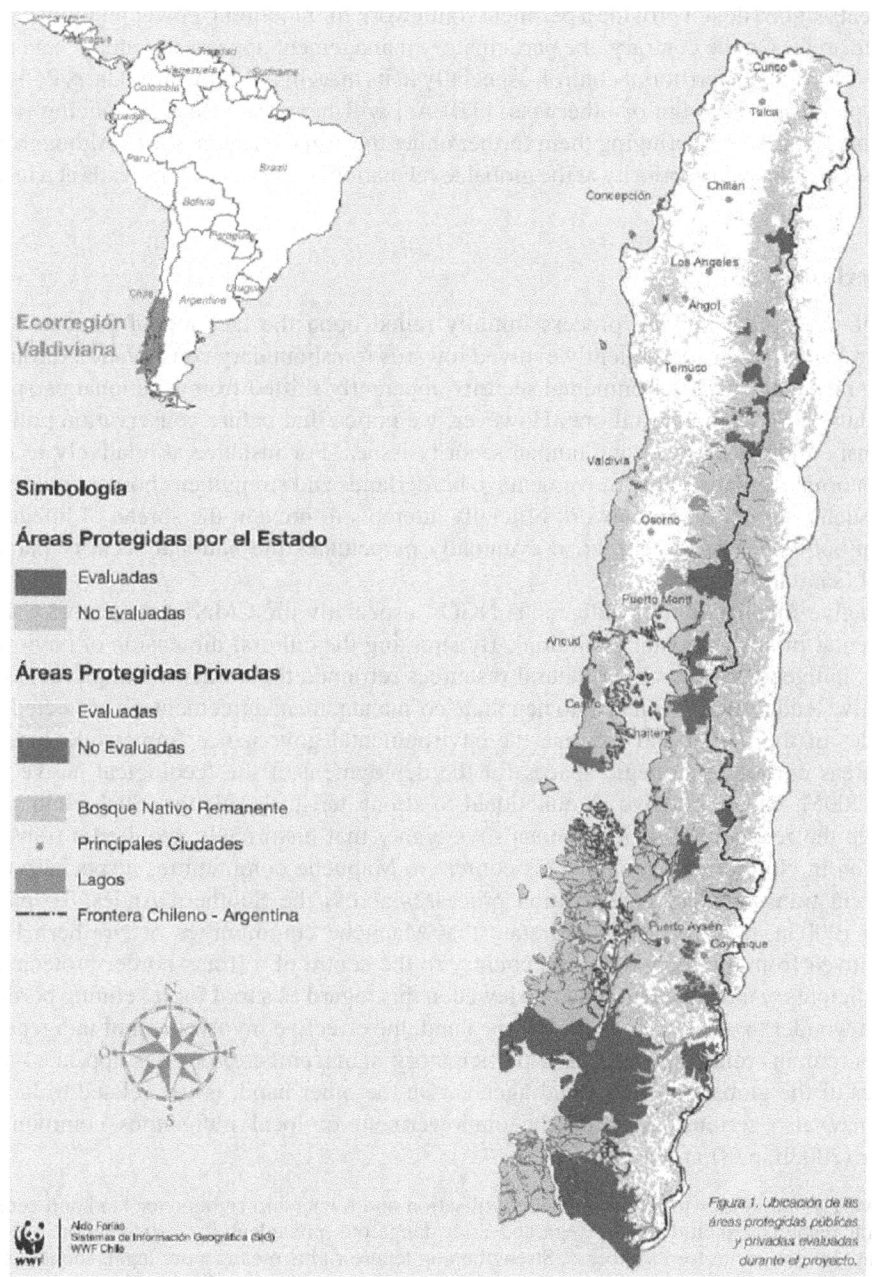

Figure 4. Valdivian Temperate Forest Ecoregion. *Source*: WWF.

However, do local indigenous communities actually benefit from this context? It appears as if the uncompleted creation of the TBRs and unresolved indigenous ecological empowerment initiatives might yield to the hegemony of global stakeholders while reinforcing their legitimacy and capacity to assist with national priorities. As highlighted above, the co-management agreements signed do not provide a pertinent framework for redefining power relationships within protected areas. On the contrary, the participatory management appears to redefine and reinforce the state's power and territorial control, especially in its margins. Wolmer (2003a, p. 265) asserted that 'in practice, by design or otherwise, [TBPAs] will have the effect of policing previously remote border areas and bringing them further under the arm of state control'. Although environmental security may be a priority at the global level, national security still prevails at a local level.

5. Conclusions

Although the border-making process initially relied upon the creation of protected areas in Northern Patagonia, it subsequently evolved towards transboundary conservation through participatory management. Environmental security apparently shifted from a national perspective to a more human or even cultural one. However, we notice that nature conservation politics simultaneously address national and human security issues. For instance, a relatively recent body of legal norms regulates cattle movements in borderlands and strengthens border control in practice. If such a normative framework officially attempts to prevent the spread of infectious diseases on both sides of the border, it eventually perpetuates the national security paradigm in terms of 'sanitary security'.

The active mobilisation of indigenous NGOs, especially the CMN in Argentina,[11] has been fundamental in this 'stagnant evolution'. By stressing the cultural dimension of environmental security, indigenous claims over natural resources reframed the border-making process. Therefore, native leaders have been able to negotiate co-management agreements in protected areas on both sides of the border and redefine the environmental governance framework. Indeed, protected areas constitute strategic spaces for the deployment of the 'ecological native' rhetoric (Ulloa, 2005), which is above all functional to strong territorial claims. Such claims, in turn, challenge the recognition of indigenous sovereignty that historically acquired a trans-Andean dimension in Northern Patagonia. This confers to Mapuche communities a special position in the current transboundary conservation processes across the Southern Andes. To paraphrase Wolmer (2003a, p. 278), we would state that Mapuche communities of Northern Patagonia 'have moved from the margins of the country to the centre of a [trans-border protected area]'.

If participatory management may be viewed in this regard as a tool for redefining borders, then we must wonder to what extent? On the one hand, the effective involvement of indigenous communities remains relatively weak and participatory management processes appear to serve the priorities of the global environmental agenda. On the other hand, issues related to land tenure rights may also seriously restrain the empowerment of local indigenous communities. As Wolmer (2003b, p. 8) argued,

> allowing communities to retain or regain utilisation and ownership rights over land and access to natural resources should not be regarded as a dangerous precedent for conservation but as an essential prerequisite for its success. Strengthening tenure rights means more legal, economic, political power for communities and greater negotiating strength in their dealings with the private sector [...].

In Chile, many Mapuche communities recently implemented private protected areas within their lands (Aylwin & Cuadra, 2011); however, such initiatives are not legally recognised, and there is no report of similar situations in Argentina. The question remains as to whether the recognition and application of the new IUCN category of Indigenous Conservation Territory will positively address this situation.

In addition, private actors are gaining crucial importance in many sectors of the region. For instance, The Conservation Land Trust foundation, which owns hundreds of thousands of hectares in Southern Patagonia, developed many conservation projects such as the Pumalín Park in the Aysén Region or, more recently, the donation of up to 40,000 ha to the Chilean State for the creation of the Yendegaia National Park in the Isla Grande de Tierra del Fuego. This foundation officially aims at '[creating] and/or [expanding] national or provincial parks to ensure the perpetuity of their ecological and evolutionary processes with the strongest long-term protection guarantee possible'.[12] Their representatives also affirm their will to promote the creation of new TBPAs. It will be interesting for future research to pay more attention to these experiences which problematise the role of private initiatives and global stakeholders in current border-making processes and reconfigurations in Patagonia.

Acknowledgements

The authors especially thank Harlan Koff and Carmen Maganda, as well as all the participants of the RISC Writer's Workshop on 'Environmental Security in Transnational contexts: What Relevance for Regional Human Security Regimes?' for their comments on a previous version of this article. We also gratefully acknowledge the additional advices of Nelson Martínez and the revision of formal aspects by Fernanda Kalazich during the preparation of the article. Comments made by the anonymous reviewers eventually helped to improve this published version.

Disclosure Statement

No potential conflict of interest was reported by the authors.

Notes

1 We use the terms 'indigenous', 'aboriginal', and 'native' interchangeably because they all refer to the first and oldest inhabitants of a specific land or territory. We therefore consider that 'indigeneity [...] connotes belonging and originariness and deeply felt processes of attachment and identification, and thus it distinguishes "natives" from others' (Merlan, 2009, p. 304). In this respect, indigeneity may be used in the context of post-colonial societies by specific groups to gain ethnic legitimacy in the reclamation of their customary territories.

2 Law n°18.362 of the Ministry of Agriculture, enacted on 8 November 1984.

3 Retrieved December 17, 2013, from http://www.ilo.org/dyn/normlex/en/f?p=NORMLEXPUB:12100:0::NO:: P12100_ILO_CODE:C169

4 Retrieved December 17, 2013, from http://www.cbd.int/convention/articles/default.shtml?a=cbd-08

5 See, for example, Mapuexpress: http://www.mapuexpress.org/ and Azkintuwe: http://www.azkintuwe.org/ websites.

6 Retrieved December 18, 2013, from http://www.mapuche-nation.org/english/main/feature/m_nation.htm

7 Resolution n°4.050, retrieved December 17, 2013, from http://intranet.iucn.org/webfiles/doc/IUCNPolicy/ Resolutions/2008_WCC_4/English/RES/res_4_050_recognition_of_indigenous_conservation_territories.pdf

8 A People's Area aiming at including some indigenous communities had been previously created by the park authority, whereas the NPA opened a 'Department of Human Settlements' in 1991 at the national level to include indigenous realities within top management.

9 This section is based on results deeply developed in a previous article published in *Territoire en Mouvement* (Sepúlveda, 2011).

10 In addition, Stoll-Kleeman, De La Vega-Leinert, and Schultz (2010) note that the Biosphere Reserve Program especially '[emphasizes] the goal of sustainable natural resource use, with community participation as a key method to achieve this', and further precise that 'the emphasis on a partnership approach is based on the conviction that BRs and their local communities are better equipped to respond to external political, economic and social pressure' (Stoll-Kleeman et al., 2010, p. 228).

11   Between 2008 and 2010, the CMN also engaged with the Chilean NGO *Observatorio Ciudadano* in comparative research on the participatory governance of protected areas in Argentina and Chile. The final report is available at http://www.observatorio.cl/sites/default/files/biblioteca/informe_final_proyecto_cmn_oc.pdf (retrieved September 24, 2013).

12   See the official website at: http://www.theconservationlandtrust.org/eng/our_mission.htm (retrieved July 18, 2014).

References

Aagesen, D. (2000). Rights to land and resources in Argentina's alerces national park. *Bulletin of Latin American Research, 19*, 547–569.

Adriazola, H., & Araya, P. (Coords.). (2007). *Reserva de Biósfera Transfronteriza Andino Norpatagónica. Documento base*. Osorno: Corporación Nacional Forestal, Gobierno de Chile.

Agnew, J., & Oslender, U. (2013). Overlapping territorialities, sovereignty in dispute: Empirical lessons from Latin America. In W. Nicholls, B. Miller, & J. Beaumont (Ed.), *Spaces of contention. Spatialities and social movements* (pp. 121–140). Surrey: Ashgate.

Amelot, X., & André-Lamat, V. (2009). La nature enfermée ou l'aire protégée comme norme de protection d'un bien commun menacé. *Géographie et Cultures, 69*, 81–96.

Anaya, S. J. (2004). International human rights and indigenous peoples: The move toward the multicultural state. *Arizona Journal of International and Comparative Law, 21*, 13–61.

Araya, P. (Ed.). (2002). *Participación de la comunidad en la gestión del Sistema Nacional de Areas Silvestres Protegidas del Estado*. Santiago de Chile: Corporación Nacional Forestal, Col. "Marco de Acción", n° 370.

Aylwin, J. (2009). Pueblo mapuche en Neuquén, Argentina, y en la Araucanía, Chile. De la fragmentación a la recon-strucción trasfronteriza. In L. Rouvière (Dir.), *Quelle(s) gouvernance(s) sur les frontières latino-américaines?* Paris: Institut pour la Gouvernance/Fondation pour le Progrès de l'homme. Retrieved from http://www.institut-gouvernance.org/fr/analyse/fiche-analyse-409.html

Aylwin, J., & Cuadra, X. (2011). *Los desafíos de la conservación en los territorios indígenas en Chile*. Temuco: Obser-vatorio de los Derechos de los Pueblos Indígenas.

Bandieri, S. (Coord.). (2001). *Cruzando la Cordillera. La frontera argentino-chilena como espacio social*. Neuquén: Universidad Nacional del Comahue, Centro de Estudios de Historia Regional.

Bandieri, S. (2009). Cuando crear una identidad nacional en los territorios patagónicos fue prioritario. *Revista Pilquen, 11*. Retrieved from http://www.scielo.org.ar/pdf/spilquen/n11/n11a11.pdf

Bello, A. (2011). *Nampülkafe. El viaje de los mapuches de la Araucanía a las pampas argentinas. Territorio, política y cultura en los siglos XIX y XX*. Temuco: Ediciones UC Temuco.

Boccara, G. (2006a). The brighter side of the indigenous renaissance. Mapuche symbolic politics and self-representation in today's Wallmapu (i.e. Chile and beyond)—Part 1. *Nuevo Mundo Mundos Nuevos*, Debates 2006. Retrieved from http://nuevomundo.revues.org/2405

Boccara, G. (2006b). The brighter side of the indigenous renaissance. Mapuche symbolic politics and self-representation in today's Wallmapu (i.e. Chile and beyond)—Part 3. *Nuevo Mundo Mundos Nuevos*, Debates 2006. Retrieved from http://nuevomundo.revues.org/2484

Cabeza, A. (1988). *Aspectos históricos de la legislación forestal vinculada a la conservación, la evolución de las áreas silvestres protegidas de la zona de Villarrica y la creación del primer parque nacional de Chile*. Santiago de Chile: Corporación Nacional Forestal, Gobierno de Chile.

Confederación Mapuche de Neuquén [CMN]. (2009). *Del co-manejo a la gobernanza en el Parque Lanín*. Neuquén: Consejo Asesor de Política Indígena Local, Documento Mapuce. Retrieved from http://www.observatorio.cl/sites/default/files/biblioteca/anexo_1.2_del_comanejo_a_la_gobernanza_en_el_lanin.pdf

Corporación Nacional Forestal de Chile [CONAF]. (1997). *Primer Congreso Latinoamericano de Parques Nacionales y otras Areas Protegidas, Informe chileno*. Santiago de Chile: Gobierno de Chile, Corporación Nacional Forestal de Chile.

Donoso, A. (2008). *Educación y nación al sur de la frontera. Organizaciones mapuche en el umbral de nuestra contem-poraneidad, 1880–1930*. Santiago de Chile: Pehuen Editores.

Fall, J. (2002). Divide and rule: Constructing human boundaries in "boundless nature". *GeoJournal, 58*, 243–251.

Fall, J. (2003). Planning protected areas across boundaries: New paradigms and old ghosts. *Journal of Sustainable Forestry, 17*(1–2), 81–102.

García, M. (2011). *Protected areas and indigenous peoples in Chile* (unpublished master's thesis). Concordia University, Montréal.

Gundermann, H., González, H., & De Ruyt, L. (2009). Migración y movilidad mapuche a la Patagonia argentina. *Magallania (Punta Arenas), 37*, 21–35.

Guyot, S. (2010). Gestion participative et populations locales dans cinq aires protégées du Chili et de l'Argentine. In J.-J. Delannoy (Ed.), *Cahiers de Géographie, 10: Espaces protégés, acceptation sociale et conflits environnementaux* (pp. 165–176). Chambéry: Université de Savoie, Collection EDYTEM.

Kent, M. (2008). The making of customary territories: Social change at the intersection of state and indigenous territorial politics on Lake Titicaca. Peru. *The Journal of Latin American and Caribbean Anthropology, 13*, 283–310.

Kradolfer, S. (2010). The transnationalisation of indigenous peoples' movements and the emergence of new indigenous elites. *International Social Science Journal, 61*, 377–388.

Mariman, P. (Comp.). (2002). *Parlamento y territorio mapuche.* Concepción: Ediciones Escaparate.

Mariman, P., Caniuqueo, S., Levil, R., & Millalen, J. (2006). *¡ . . . Escucha, winka . . . ! Cuatro ensayos de Historia Nacional Mapuche y un epílogo sobre el futuro.* Santiago de Chile: LOM Ediciones.

Merlan, F. (2009). Indigeneity: Global and local. *Current Anthropology, 50*, 303–333.

Miniconi, R., & Guyot, S. (2010). Conflicts and cooperation in the mountainous Mapuche territory (Argentina). The case of the Nahuel Huapi National Park. *Journal of Alpine Research, 98*, 123–140.

Molina, R. (2009). Relaciones transfronterizas entre atacameños y collas en la frontera norte chilena-argentina. La desintegración de espacios y articulaciones tradicionales indígenas. In L. Rouvière (Dir.), *Quelle(s) gouvernance(s) sur les frontières latino-américaines?* Paris: Institut pour la Gouvernance/Fondation pour le Progrès de l'homme. Retrieved from http://www.institut-gouvernance.org/fr/analyse/fiche-analyse-408.html

Muller, S. (2003). Towards decolonisation of Australia's protected area management: The Nantawarrina Indigenous Protected Area experience. *Australian Geographical Studies, 41*, 29–43.

Nahuel, H. J. (2008). *Foro de pueblos indígenas, comunidades campesinas y afrodescendientes.* Buenos Aires: Secretaría de Ambiente y Desarrollo Sustentable de la Nación.

Navarro Floria, P. (2008). La construction des territoires nationaux latino-américains vue depuis leurs marges. *Les Cahiers ALHIM, 16*, 167–180.

Navarro Floria, P., & Delrio, W. (Comps.). (2011). *Cultura y espacio: Araucanía—Norpatagonia.* San Carlos de Bariloche: Universidad Nacional de Río Negro, Instituto de Investigaciones en Diversidad Cultural y Procesos de Cambio.

Núñez, A. (2013). La frontera no deja ver la montaña: invisibilización de la cordillera de Los Andes en la Norpatagonia chileno-argentina. *Revista de Geografía Norte Grande, 55*, 89–108.

Núñez, A., Sánchez, R., & Arenas, F. (Eds.). (2013). *Fronteras en movimiento e imaginarios geográficos. La cordillera de Los Andes como espacio socio-cultural.* Santiago de Chile: RIL Editores.

Núñez, P., Matossian, B., & Vejsbjerg, L. (2012). Patagonia, de margen exótico a periferia turística. Una mirada sobre un área natural protegida de frontera. *Pasos, 10*, 47–59.

Pérez Raventós, A., & Biondo, C. (2003). *Una nueva relación en el Parque Nacional Lanín: el comité de gestión del comanejo con "determinación" mapuche* (Claspo Research report on Proyecto Comparado sobre Políticas Públicas). Retrieved from http://lanic.utexas.edu/project/laoap/claspo/rtc/0011.pdf

Pinto, J. (Ed.). (1996). *Araucanía y pampas, un mundo fronterizo en América del Sur.* Temuco: Ediciones Universidad de la Frontera.

Pinto, J. (Ed.). (2011). *Araucanía, siglos XIX y XX. Economía, migraciones y marginalidad.* Osorno: Editorial Universidad de Los Lagos.

Ramousse, D., & Salin, E. (2007). Aires protégées des périphéries sud-américaines: entre réserves stratégiques et valorisation patrimoniale. *Mondes en développement, 138*, 11–26.

Sandwith, T., Shine, C., Hamilton, L., & Sheppard, D. (2001). *Transboundary Protected Areas for Peace and co-operation.* Gland: IUCN/Cardiff University, Best Practice Protected Area Guidelines Series, 7.

Sepúlveda, B. (2011). Pentecôtisme et recompositions territoriales autochtones dans le Chili central. Le cas des communautés pehuenches de Lonquimay. *Territoire en Mouvement, 13*, 84–101.

Sepúlveda, B. (2012). Gestion participative en territoires autochtones: disputes autour d'une aire protégée dans les Andes chiliennes. *Cahiers de Géographie du Québec, 56*, 621–639.

Stoll-Kleeman, S., De La Vega-Leinert, A. C., & Schultz, L. (2010). The role of community participation in the effectiveness of UNESCO Biosphere Reserve management: evidence and reflections from two parallel global surveys. *Environmental Conservation, 37*, 227–238.

Ulloa, A. (2005). *The ecological native: Indigenous peoples' movements and eco-governmentality in Columbia.* London: Routledge.

Valenzuela, I., & Contreras, J. P. (2000). *Conservación y desarrollo indígena: hacia una vinculación positiva en áreas silvestres protegidas.* Antofagasta: Ministerio de Agricultura, Corporación Nacional Forestal de Chile.

Wolmer, W. (2003a). Transboundary conservation: The politics of ecological integrity in the Great Limpopo Transfrontier Park. *Journal of Southern African Studies, 29*, 261–278.

Wolmer, W. (2003b). *Transboundary Protected Area governance: Tensions and paradoxes*. Paper presented at the workshop on Transboundary Protected Areas in the Governance Stream of the 5th World Parks Congress, Durban, South Africa. Retrieved from http://www.tbpa.net/docs/WPCGovernance/WilliamWolmer.pdf

Zbicz, D. (2003). Imposing transboundary conservation: Cooperation between internationally adjoining protected areas. *Journal of Sustainable Forestry, 17*(1–2), 21–37.

Index

Note: Page numbers in *italic* type refer to figures
Page numbers in **bold** type refer to tables
Page numbers followed by 'n' refer to notes